MODERN TECHNIQUES IN GEOGRAPHY
General Editor: Joan Garlick, M.A.

SAMPLING TECHNIQUES IN GEOGRAPHY

Roger Dalton
Joan Garlick
Roger Minshull
Alan Robinson

LONDON : GEORGE PHILIP AND SON LIMITED

THE AUTHORS

Roger Dalton, M.A., Head of the Department of Geography, Bishop Lonsdale College, Derby.

Joan Garlick, M.A., Head of the Department of Geography, Bishop Grosseteste College, Lincoln.

Roger Minshull, M.A., Senior Lecturer in American Studies, Bishop Grosseteste College, Lincoln.

Alan Robinson, M.A., Lecturer in Community Studies, New University of Ulster.

© 1975 ROGER DALTON JOAN GARLICK ROGER MINSHULL ALAN ROBINSON

All rights reserved. No part of this publication may be reproduced, stored in a retrieval system, or transmitted, in any form or by any means, electronic, mechanical, photocopying, recording or otherwise, without prior permission. Enquiries should be addressed to the publishers, George Philip and Son Limited, 12–14 Long Acre, London WC2E 9LP.

Set by H Charlesworth and Co Ltd Huddersfield, and printed in Great Britain by J. W. Arrowsmith Ltd, Bristol. ISBN 0 540 00984 9

Contents

	Preface	iv
	Introduction	1
I	**AREAL SAMPLING FROM MAPS**	
1	Procedures for selecting the sample	3
2	Systematic point sampling	13
3	Random point sampling	26
4	Stratified point sampling	30
5	Systematic linear sampling and other methods	34
II	**NON-AREAL METHODS OF SAMPLING**	
6	Sampling procedures using interval data	36
7	Field sampling from an unknown population total	47
8	Household sampling with uniform sampling fraction	50
9	Household sampling with variable sampling fraction	57
10	Sampling in a study of farm economy	60
11	Stratified sampling in settlement study	69
III	**COMPARISON OF SAMPLE VALUES**	
12	Significance of sample differences for interval data	79
13	Significance of sample differences for frequency data	84
	Conclusion	87
	Selected Bibliography	88
	Appendices	89

Preface

This third volume in the series *Modern Techniques in Geography* is concerned with sampling procedures appropriate to work using maps and field surveys. We feel that this is a most important topic for those geographers who seek for greater accuracy in their work and a hypothesis-testing or problem-solving approach to geographical understanding. In both these circumstances it is essential that the procedure for collecting data should be properly constructed and the evaluation of that data undertaken with a clear knowledge of its limitations.

We trust that in this volume we have answered the following questions. What are the various sampling techniques available? How can an appropriate technique be applied to work involving either map analysis or field survey? What are the limitations to the value of the results obtained? How can comparisons be made between sets of sample data?

In structuring the book it has not proved feasible to answer all these questions for each example when it is first introduced. Techniques are explained, then examples of their usage given. The same data may be used later to illustrate a further stage of analysis.

We have purposely included a good deal of computation. This is not because the techniques are mathematically difficult but because we feel that an introductory volume such as this should make each stage of the working clear. Cutting corners is a sure way of spreading confusion and making errors.

Calculations throughout have been completed correct to one place of decimals except in a few instances, where it will be stated or apparent.

We are indebted to the Literary Executor of the late Sir Ronald A. Fisher, F.R.S., to Dr. Frank Yates, F.R.S., and to Longman Group Ltd., London, for permission to reprint the table on page 92 from their book *Statistical Tables for Biological, Agricultural and Medical Research*.

R.T.D.   E.J.G.   R.M.M.   A.R.

Introduction

Sampling is so common a practice that most people have had some experience of it. It is a method of estimating some property of a large population by testing a small part of it. The word 'population' is used in sampling to mean the whole of the phenomenon being investigated and the term is applied to both living and inanimate things.

In geographical study there are many phenomena which cannot be studied in entirety and sampling is an answer to the coverage problem when an area is too large, or contains too much data to be investigated in the time and with the manpower available. The purpose of sampling is therefore to enable valid generalisations to be made without studying every example or member of a population. Accordingly, it is necessary to investigate methods of sampling in order that the correct one may be chosen for the job in hand and also that the method yields as accurate a generalisation as possible.

A clear distinction must be made between two very different concepts of sampling.

Purposive sampling

So called 'sample studies' are examples of this. A more accurate name for these is 'case studies'. In this type a discrete place such as a farm, village, town, or piece of landscape may be selected for study because it is thought to be typical, in some fundamental aspects, of conditions over a wider area. The choice of unit is dependent on the availability of material and on the skill of the selector[1]. Most teachers have used purposive sampling in field work. Such types of sample studies have a place in geography teaching and much good classroom work has been accomplished by using them.

Probability sampling

It is this type of sampling which is to be considered in this book. The choice of the individual cases to be studied is based on strict mathematical theory and the cases are selected according to established rules, so that the characteristics of a population can be ascertained with a stated degree of confidence from the information established by studying the sample cases. It is possible to calculate what the probabilities are that the information derived from a sample can be

---

[1] For examples of the use of different units see:
　　Widdup, H. L., and R. C. Honeybone, (eds.), *Sample Studies*, Geographical Association, 1962.
　　Dempster, P., *Asian Sample Studies*, Geographical Association, 1968.
　　Highsmith, R. M. *et al*, *Case Studies in World Geography*, Prentice-Hall, 1964.
　　Young, I. V., *Farm Studies in Schools*, Association of Agriculture, 1968.

accurately applied to the whole. The sample can reproduce in miniature the characteristics of the area or topic under investigation with a known degree of accuracy.

While purposive sampling has long attracted geographers, probability sampling was not generally recognised as something which might be of value to their discipline until the 1950s. Sociologists were among the first to realise that a randomly selected sample of households, for example, would exhibit collectively the same characteristics as the total households in an area.

Probability sampling is now regarded as being more efficient in time, accuracy and objectivity than a necessarily incomplete attempt at total analysis. It is increasingly used in the collection of data for hypothesis testing.

For valid generalisations to be made statistical rules must be strictly followed and in this book an attempt is made to explain these rules simply. Detailed examples of a selection of sampling methods for differing types of data are set out together with calculations necessary to enable the results to be applied to the whole population. The practical problems of implementing the methods are also discussed. This is a vital consideration as the putting of theory into practice is not nearly as straightforward as might first appear, especially when collecting data in the field.

I — AREAL SAMPLING FROM MAPS

CHAPTER ONE

Procedures for selecting the sample

Areal sampling methods may be used for both sampling areal distributions with continuous cover such as types of land use or types of landform and for selecting places for study, such as farms or villages, from a map. The latter may also be done non-areally from lists and will be dealt with in Part II of this book.

Sampling units

For areal sampling there are three main sampling units: points; lines or traverses; quadrats or areas (*Figure 1*).

1) The *point* is the simplest element consisting of a dot on the map or on the ground. The point may be identified, that is, its location stated by its grid coordinates. At the point identified some characteristic is either present or absent. For example the land use at a given point is either arable or not arable, wooded or not wooded. The feature being sampled, in this case land use, may be categorised in an appropriate way and the relevant category identified. Figure 1 column (i) shows three possible arrangements of selected sample point units.
2) *Lines or traverses* can be used as sample units to select information from maps or in the field. It is the proportion of the length of the line occupied by some particular feature, such as land use type, which is recorded as the sample data. Figure 1 column (ii) shows three possible arrangements of sample lines.
3) *Quadrats* are much used by ecologists and botanists in the field and are of particular use in geographical sampling of vegetation. The quadrat unit is not as easy to handle as the line as it is more difficult to measure the proportion of the area occupied by the feature being investigated than it is to measure its occurrence along a line. Figure 1 column (iii) shows three possible arrangements of sampling quadrats.

Figure 1 also indicates that not only is there a choice of unit for sampling, but also different procedures for selecting the units which are to comprise the sample. The main types of procedure are random and systematic and these may be used singly or in combination.

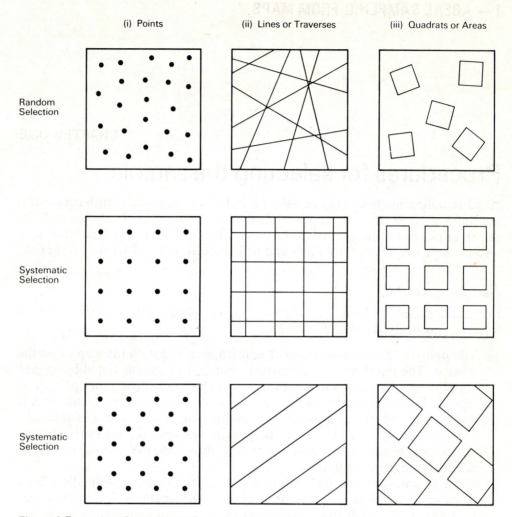

Figure 1 Examples of sampling units and methods of sampling

Random sampling procedures

As the object is to secure a sample which will reproduce the characteristics of the total population as closely as possible, the main requirement is to avoid bias in the selection of units. To this end it must be ensured that every part of the area has an equal chance of being selected and no deliberate choosing of any unit which looks as if it is 'average' must be allowed. This would give rise to error due to bias and this can be eliminated by sticking strictly to the rules for random selection. There is still the sampling error due to chance differences between the members of the population included in the sample and those not included. This error can be calculated from the data collected in the sample provided the sample is of sufficient size and is unbiased in the initial stage of unit selection. The problem of size of sample will be demonstrated in the examples worked in Chapters 2, 3, 4 and 5.

Unbiased random selection of points is obtained by use of either 20-sided[1] dice or from a random numbers table (*Appendix 2*). The 20-sided die has figures 0–9, each figure appearing on two of the 20 sides. If up to 100 points are required two dice of different colours are necessary, one colour to represent tens and one to represent units. The two dice are thrown together. If the tens die shows 5, for example, and the units die 6, the random number will be 56. If the tens die shows 0 and the units die 6, the random number will be 6. If up to 1000 points are required in the sample three dice of different colour must be used. A table of random numbers may be compiled by listing in order the numbers resulting from throwing the dice the required number of times. Alternatively printed random numbers tables can be used, such as in the example in Chapter 3. The numbers must be used strictly in sequence either horizontally or vertically. A start may be made anywhere on the table and the direction of using the sequence must remain constant. It is advisable to cross off the numbers as they are used. For example in Appendix 2 start at the top of the left-hand column and work down; if up to 100 numbers are required (that is if there are 100 points on the map from which to select the sample) use two columns of digits, the left hand column representing tens and the right hand column units. The first number would thus be 20, the second 74, the third 94 and so on. When there are less than 100 points on the map from which to select a sample, 60 for example, numbers between 60 and 100 are of no use and should be discarded. If there are 50 points the series can be re-phased by taking 50 from each number greater than 50 which arises, e.g. 51 would become 1, 65 would become 15. Each point would then have two chances of being selected. If there are up to 1000 points on the map grid from which to select a sample three columns of digits must be used. It is important that numbers are taken strictly in sequence.

To use the numbers for the selection of sample points on the map a grid must be put over the map and coordinates numbered, starting in the south-west corner. When using Ordnance Survey maps the National Grid printed on the map can be used. This method of selecting random points is demonstrated in the example in Chapter 3.

Random lines can be selected in a number of ways using the random numbers table:

1) Use two digit random numbers to locate the starting point of either horizontal or vertical grid lines. This would be a random sample of grid lines. e.g. Random number 26 would refer to point 800326 if horizontal

---

[1] Available from E. J. Arnold and Son Ltd., Leeds.

lines were being used. If vertical lines were to be used it would refer to point 826300.

2) Let a four digit random number represent two pairs of eastings, the first pair to be located along the southern edge of the map and the second pair along the northern edge. e.g. The line plotted from the random number 2017 would run from grid reference 820300 to 817400. The lines located thus would all run between the northern and southern edges of the map, or alternatively between the eastern and western edges if the pairs of numbers represented northings.

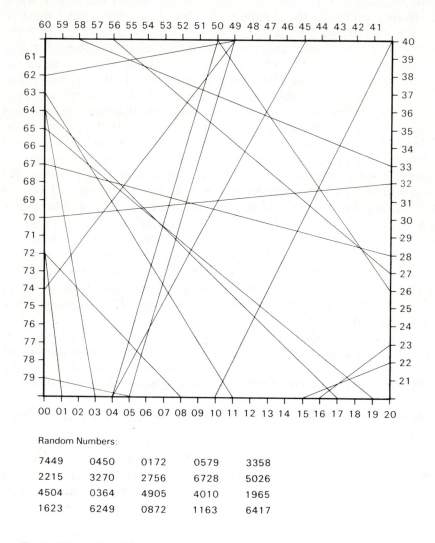

Random Numbers:

7449	0450	0172	0579	3358
2215	3270	2756	6728	5026
4504	0364	4905	4010	1965
1623	6249	0872	1163	6417

Figure 2 Location of random lines — method 3

3) Place a grid of 0.5 kilometre squares over the map. Starting in the south-west corner use numbers 00–19 along the southern edge, numbers 20–39 along the eastern edge, numbers 40–59 along the northern edge and numbers 60–79 along the western edge, as shown in Figure 2. The lines can be located according to their starting and ending points on the edges of the map. The starting point is located by the first pair of digits and the ending by the second pair in each four digit random number. e.g. The line located by random number 7449 would run diagonally from the western edge to the northern edge. Lines would be of unequal length and would run in various directions giving a strictly random coverage (*Figure 2*).

Care should always be taken to use numbers in sequence and to list sufficient numbers before locating them all on the map. It will be noticed that the numbers of the grid end at 79. In the table of random numbers there will be pairs of numbers greater than this, e.g. the third number down in the left hand column is 9470 (*Appendix 2*). Both these pairs must be ignored. Some combinations of pairs of digits in each four digit random number will produce lines coincident with the frame of the grid, e.g. 2017. It must be decided beforehand whether or not such lines will be included in the sample. Other combinations of pairs of digits will produce the same line or part of the same line on the grid, e.g. 7449 and 4974; 6100 and 6400. The second number must either be ignored or the data recorded twice for the duplicated line. Whatever line of action is adopted it is important to be consistent to maintain the random nature of the sample.

Systematic sampling procedures

Systematic aligned

This is a method whereby only the first individual unit is selected at random. All subsequent units in the sample are chosen at a prescribed and uniform interval from the first. An even set of sample points with regular spacing both ways results. This method is speedier and simpler to use than the random method. It has the disadvantage however that all parts of the area do not have an equal chance of selection once the first point has been established. The initial choice determines the rest. Also care is needed to ensure that there is not some recurring periodic feature in the distribution which might be picked out repeatedly by the evenly spaced points, lines or quadrats and so give rise to bias. The investigator must judge whether or not material is suitable for systematic sampling.

Systematic unaligned

The last named disadvantage may be lessened considerably by use of the unaligned method which combines systematic and random procedures. The selection of the sample points is as follows and is illustrated by Figure 3.

PROCEDURES FOR SELECTING THE SAMPLE

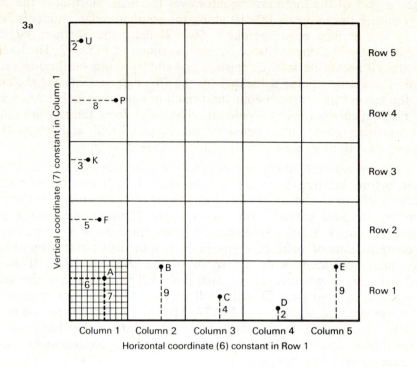

Figure 3 Systematic unaligned sampling

1) Divide the study area by a regular grid into sub-areas.
2) Divide each sub-area by a 10 × 10 grid, e.g. in Figure 3a, the south-west sub-area.
3) Locate the first sample point A in the south-west sub-area by a pair of random numbers selected either by throwing two 20 sided dice, or from a random numbers table. In this case the numbers are 6 and 7. The 6 is the horizontal and the 7 the vertical coordinate.
4) Along the first row the horizontal coordinate is kept constant (in this case 6) and the vertical coordinate is varied by use of a series of four random numbers obtained either by throwing one die, or in sequence from the same column of the random numbers table, giving point B, C, D and E. If these random numbers are 9, 4, 2 and 9, then B = 69, C = 64, D = 62, E = 69.
5) On the first column the vertical coordinate of A, which is 7, remains constant and the horizontal coordinate is varied by use of the next four random numbers (5, 3, 8, 2), giving points, F, K, P and U in the first column. Thus F = 57, K = 37, P = 87, U = 27.
6) A new point G is located using the horizontal coordinate of F and the vertical coordinate of B; F and B are the first points in the second row and column. The coordinates of G are thus 59.
7) Complete the second row using the horizontal coordinate 5 and vary the vertical coordinate by continuing to use the random numbers table or by throwing the die.
8) Complete the second column in the same way using the vertical coordinate 9 and varying the horizontal coordinate.
9) A new corner point M has the horizontal coordinate of L (3) and the vertical coordinate of H (2); M is therefore 32. Continue in this way to complete the frame (*Figure 3b*). This method gives a set of points with irregular spacing in one direction at a time.

Two variations of random and systematic sampling of particular use in geographical studies may now be considered. If the phenomenon being investigated is unevenly distributed or if the study area is particularly large, making adequate coverage of the area impracticable using the above methods, some initial grouping of the area can be an advantage. Stratified sampling is in fact essential in studies of areas with distinct sub-regions whilst cluster sampling is useful in areas too extensive for overall sampling.

Stratified sampling procedures

This method of sampling involves the selection of individual points either randomly or systematically within pre-arranged sub-areas (*Figure 4*), which is often advisable when a heterogeneous area is being investigated. This means

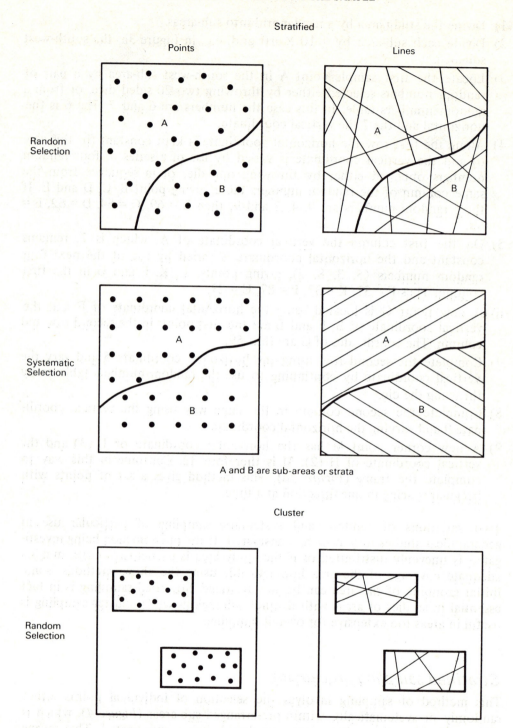

Figure 4 Examples of stratified and cluster sampling frames

that the area should be divided into smaller sub-areas which are more homogeneous in respect of some attribute which affects the distribution of the characteristic being investigated. Such areas require smaller samples than heterogeneous areas. The study area would thus be divided purposively into natural segments according to some stated criterion, e.g. according to geological characteristics or relief characteristics or a combination of both. e.g. Division may be made into sub-areas of clay lowland and limestone upland if these were relevant to the particular study. Likewise, in an urban area, division may be according to some predetermined city zonation, such as residential, industrial and commercial areas. Independent samples would then be drawn from each sub-division; the number of individuals selected from each sub-area can be proportional to the size of the area. e.g. If 100 units are wanted in the sample in all and 40% of the area is upland and 60% lowland, 40 can be selected from the upland and 60 from the lowland. If systematic procedure is being followed then proportional representation occurs automatically.

Stratified sampling may also be desirable, for example, if one wished to select about 30 of the urban administrative areas in England and Wales for study as a sample of all such areas. To do this areally England and Wales would first be divided into regions according to some suitable stated criterion, depending on the purpose of the study. An appropriate breakdown might be into Economic Planning Regions. Alternatively some general characteristic may be taken as the criterion for division. In this case national average unemployment would be appropriate with separation of those areas with above average unemployment from those with below average unemployment. In the former case four urban administrative areas could be selected randomly from each of the eight Economic Planning Regions; in the latter, fifteen from each of the two sub-regions according to unemployment rate. This would ensure a more even spatial coverage than if England and Wales were sampled as a whole.

Cluster sampling procedures

This approach can be used in similar situations but in this method samples are selected from sub-areas determined randomly (*Figure 4*), not purposively as for stratified sampling.

In all a great variety of types of areal sampling is possible owing to the various combinations of different sampling units with different sampling procedures. Nine of these were shown on Figure 1.

The methods of selecting points, lines or quadrats for sampling areal distributions with continuous cover can be adapted for use in selecting villages, farms, streets or houses in a town. The distribution of these is not continuous and the selection of places from a map is more difficult if their distribution is irregular. If they are clustered and large 'empty' spaces occur between these groups then choosing a sample will be easier from a list rather than from a map. However the areal method can be used, as follows. A grid is placed over the map of the area chosen for study and on it is marked the requisitive number of randomly or systematically selected points. (A few extra points should be added since some may turn out to be unusable.) The settlements to be included in the sample are those on or nearest the sample points. To make selection consistent, only those villages nearest to the sample points in a specified direction are included. If settlement is close a fine grid is necessary or bias may be introduced, with some settlements having no chance of selection. The grid, if drawn out on tracing paper larger than the map could be superimposed on the map by the selection of a random number with which to locate its centre, in an effort to avoid bias.

CHAPTER TWO

Systematic point sampling

To demonstrate the techniques of areal sampling in detail examples will be worked in this and the following three chapters using different sampling procedures. The same data will be used throughout, although samples of different sizes will be taken and the results compared. The results will be processed in each case and the methods of calculating sampling error will be demonstrated. The accuracy with which results can be applied to the whole area will be considered together with the question of choosing the size of sample necessary to get adequate results. The efficiency and disadvantages of each method will become apparent.

Although the random method of unit selection has been described first as this is the method which best eliminates bias, in the examples following systematic methods will be demonstrated first as they are easier to handle. The 1: 25 000 Land Use Sheet 838 Coldstream is used and the area of 100 square kilometres enclosed by northings 30 and 40, and eastings 80 and 90 is chosen for study. The area has few scattered settlements, a small amount of scattered woodland, a considerable proportion of rough land unevenly distributed over the area, and the bulk of the area is divided between pasture and arable. The problem is to estimate the amount of land taken up by the various land uses. The results obtained from the four sampling techniques will be compared.

Sample of 100 points

The one kilometre grid lines printed on the map are used to locate the sample points. Taking the intersections of the grid lines, but excluding those on northing 40 and easting 90, we have 100 points in all (*Figure 5*). The land use at each point is recorded on Table 1 and the frequency of occurrence of each category of land use is tabulated.

It is tempting to take the information so derived at its face value, as if the percentage occurrence had actually been measured, but this would be very dangerous. In order to generalise from this sample data, i.e. to estimate from it the percentage of arable over the whole of the area we must calculate what the probabilities are that the percentages derived from the sample survey can be applied to the whole area being studied. The method of estimating the reliability of the facts involves calculation of the standard error for the frequency of each land use category in turn, or every category in which the worker is interested. As this is a point survey we have not measured actual amounts of land under various usages; we have only collected frequency data, e.g. arable occurred 32 times and grassland 52 times, out of 100 possible occurrences.

As the sample data gives 32% arable this implies that 68% is not arable.

SYSTEMATIC POINT SAMPLING

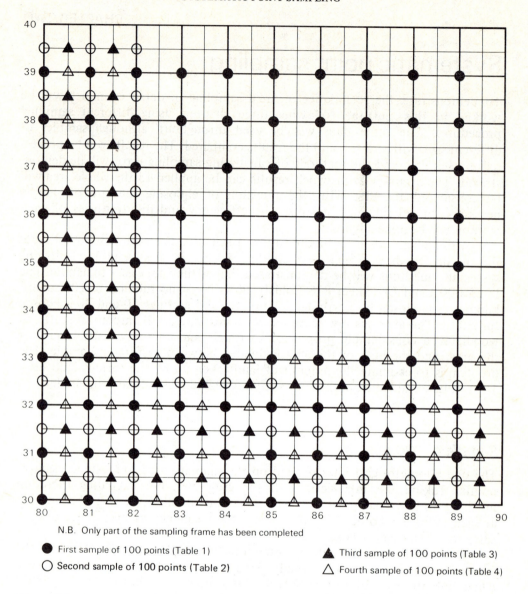

● First sample of 100 points (Table 1)
○ Second sample of 100 points (Table 2)
▲ Third sample of 100 points (Table 3)
△ Fourth sample of 100 points (Table 4)

Figure 5 Systematic areal sampling — location of sample points

These two proportions together equal the total size of the sample and thus the total number of possibilities of land use being arable.

The relationship of the true proportions to these sample proportions will depend upon the size of the sample. In other words the size of the sample will affect the standard or average error of the sample value. This sampling error is therefore the difference between estimates made from the sample and the correct value. The larger the sample the smaller the standard error will be as it varies inversely with the size of the sample. The formula for the standard error for binomial data (i.e. data comprising two mutually exclusive numbers or sets of conditions) is:

$$SE = \sqrt{\frac{p \times q}{n}}$$

where SE = the standard error
        p = the percentage of land in a certain category
        q = the percentage of land *not* in that category
        n = the number of points in the sample

Table 1 First systematic sample of 100 points

Point	Land use	Point	Land use	Point	Land use	Point	Land use	Point	Land use
8030	MG	8031	G	8032	A	8033	G	8034	A
8130	G	8131	G	8132	WTR	8133	G	8134	MG
8230	G	8231	W	8232	G	8233	G	8234	A
8330	G	8331	G	8332	A	8333	A	8334	G
8430	A	8431	G	8432	G	8433	G	8434	G
8530	A	8531	R	8532	R	8533	G	8534	G
8630	G	8631	R	8632	R	8633	G	8634	G
8730	R	8731	G	8732	G	8733	G	8734	G
8830	R	8831	R	8832	A	8833	G	8834	A
8930	G	8931	R	8932	R	8933	A	8934	G
8035	A	8036	A	8037	G	8038	A	8039	G
8135	A	8136	S	8137	G	8138	A	8139	G
8235	G	8236	G	8237	A	8238	A	8239	A
8335	G	8336	A	8337	A	8338	A	8339	A
8435	G	8436	A	8437	A	8438	G	8439	WTR
8535	G	8536	A	8537	A	8538	G	8539	A
8635	G	8636	G	8637	G	8638	G	8639	G
8735	G	8736	G	8737	A	8738	A	8739	G
8835	G	8836	G	8837	G	8838	A	8839	A
8935	W	8936	G	8937	G	8938	G	8939	A

A Arable   MG Market gardening   S Settlement   WTR Water
G Grassland   R Rough land   W Woodland

NUMBER OF OCCURRENCES

A	G	R	MG	W	WTR	S	Total
32	52	9	2	2	2	1	**100**

As the total number of points is 100 these figures represent percentages. They show the percentage frequency of occurrence of each type of land use.

Arable:
- p = percentage of sample in arable = 32%
- q = percentage of sample not in arable = 68%
- n = number of points in sample = 100

Substituting the sample values for p, q, and n in the formula:

$$SE = \sqrt{\frac{32 \times 68}{100}} = \sqrt{21.8} = 4.7$$

According to the laws of probability[1] there is a 68% chance that the true amount of arable will lie within one standard error of the sample amount. This gives a range of values from 32.0−4.7 to 32.0 + 4.7, i.e. 27.3 to 36.7% at the 68% confidence level. This is a low level of probability and does not help much in assessing the land use of the area.

At the 95% confidence level, i.e. a level of probability of 95%, we can say that the true percentage of arable lies within 2 standard errors of the sample amount and at the 99% confidence level within 3 standard errors of the sample amount. Thus at the 95% confidence level the true value will lie between 32.0 ± 2SE, i.e. between 22.6 and 41.4%.

At the 99% confidence level it will lie between 32.0 ± 3SE, i.e. between 17.9 and 46.1%.

For the purpose of these exercises the 95% confidence level will be used.

Grassland:
- p = percentage of sample in grassland = 52%
- q = percentage of sample not in grassland = 48%
- n = number of points in sample = 100

$$\therefore SE = \sqrt{\frac{52 \times 48}{100}} = \sqrt{25.0} = 5.0$$

Thus at the 95% confidence level the true value for grassland will lie between 52.0 ± 2SE, i.e. between 42.0 and 62.0%.

Rough land:

$$p = 9\%, q = 91\%, n = 100$$

$$\therefore SE = \sqrt{\frac{9 \times 91}{100}} = \sqrt{8.2} = 2.9$$

Thus at the 95% confidence level the true value for rough land will lie between 9.0 ± 2SE, i.e. between 3.2 and 14.8%.

It can be seen that with a sample of only 100 points we cannot make very precise statements about the true amount of land under different land uses.

1 For further details see:
Cole, J. P., and C. A. M. King, *Quantitative Geography*, Wiley, 1968
Gregory, S., *Statistical Methods and the Geographer*, Longman, 1973

This point is underlined by taking a second, third and fourth independent sample of 100 systematically selected points. The land use is recorded on Tables 2, 3 and 4 and the standard errors and range of population values are calculated for each sample. The sampling frame for these, i.e. the location of the points, is shown in Figure 5.

Table 2 Second systematic sample of 100 points

Point	Land use	Point	Land use	Point	Land use	Point	Land use	Point	Land use
800305	W	800315	A	800325	A	800335	G	800345	A
810305	W	810315	G	810325	A	810335	G	810345	A
820305	A	820315	G	820325	G	820335	A	820345	MG
830305	G	830315	A	830325	A	830335	G	830345	A
840305	G	840315	G	840325	G	840335	G	840345	W
850305	G	850315	W	850325	G	850335	S	850345	G
860305	G	860315	R	860325	R	860335	A	860345	G
870305	G	870315	G	870325	G	870335	G	870345	G
880305	R	880315	R	880325	A	880335	G	880345	A
890305	R	890315	R	890325	W	890335	G	890345	G
800355	A	800365	A	800375	G	800385	G	800395	G
810355	A	810365	G	810375	G	810385	G	810395	A
820355	A	820365	S	820375	G	820385	G	820395	A
830355	G	830365	A	830375	A	830385	MG	830395	G
840355	G	840365	W	840375	A	840385	A	840395	G
850355	A	850365	G	850375	G	850385	WTR	850395	A
860355	MG	860365	G	860375	A	860385	G	860395	G
870355	A	870365	G	870375	A	870385	G	870395	A
880355	G	880365	MG	880375	A	880385	G	880395	G
890355	G	890365	G	890375	G	890385	A	890395	G

NUMBER OF OCCURRENCES

A	G	R	MG	W	WTR	S	Total
31	50	6	4	6	1	2	**100**

Arable:
$$SE = \sqrt{\frac{31 \times 69}{100}} = \sqrt{21.4} = 4.6$$

Range of values at the 95% confidence level: 21.8 to 40.2%

Grassland:
$$SE = \sqrt{\frac{50 \times 50}{100}} = \sqrt{25.0} = 5.0$$

Range of values at the 95% confidence level: 40.0 to 60.0%

Rough land:
$$SE = \sqrt{\frac{6 \times 94}{100}} = \sqrt{5.6} = 2.4$$

Range of values at the 95% confidence level: 1.2 to 10.8%

Table 3 Third systematic sample of 100 points

Point	Land use	Point	Land use	Point	Land use	Point	Land use	Point	Land use
805305	G	805315	R	805325	G	805335	MG	805345	G
815305	G	815315	G	815325	G	815335	A	815345	G
825305	G	825315	G	825325	G	825335	G	825345	A
835305	G	835315	A	835325	A	835335	G	835345	G
845305	G	845315	G	845325	G	845335	G	845345	A
855305	R	855315	WTR	855325	G	855335	G	855345	G
865305	G	865315	R	865325	G	865335	G	865345	G
875305	R	875315	G	875325	A	875335	G	875345	G
885305	R	885315	R	885325	S	885335	S	885345	G
895305	W	895315	G	895325	R	895335	G	895345	W
805355	A	805365	A	805375	A	805385	G	805395	A
815355	A	815365	A	815375	A	815385	G	815395	G
825355	A	825365	A	825375	A	825385	G	825395	MG
835355	G	835365	A	935375	G	835385	G	835395	W
845355	G	845365	MG	845375	G	845385	G	845395	A
855355	A	855365	A	855375	A	855385	A	855395	G
865355	G	865365	A	865375	W	865385	A	865395	A
875355	R	875365	G	875375	A	875385	W	875395	A
885355	G	885365	G	885375	G	885385	G	885395	A
895355	G	895365	G	895375	S	895385	A	895395	A

NUMBER OF OCCURRENCES

A	G	R	MG	W	WTR	S	Total
30	50	8	3	5	1	3	100

Arable:
$$SE = \sqrt{\frac{30 \times 70}{100}} = \sqrt{21.0} = 4.6$$

Range of values at the 95% confidence level: 20.8 to 39.2%

Grassland:
$$SE = \sqrt{\frac{50 \times 50}{100}} = \sqrt{25.0} = 5.0$$

Range of values at the 95% confidence level: 40.0 to 60.0%

Rough land:
$$SE = \sqrt{\frac{8 \times 92}{100}} = \sqrt{7.4} = 2.7$$

Range of values at the 95% confidence level: 2.6 to 13.4%

Table 4 Fourth systematic sample of 100 points

Point	Land use	Point	Land use	Point	Land use	Point	Land use	Point	Land use
805300	G	805310	R	805320	G	805330	A	805340	G
815300	A	815310	G	815320	G	815330	W	815340	A
825300	G	825310	G	825320	R	825330	G	825340	R
835300	G	835310	R	835320	G	835330	G	835340	G
845300	G	845310	G	845320	G	845330	G	345340	A
855300	G	855310	G	855320	R	855330	G	855340	S
865300	R	865310	G	865320	G	865330	G	865340	S
875300	R	875310	R	875320	G	875330	G	875340	G
885300	R	885310	R	885320	R	885330	G	885340	G
895300	G	895310	G	895320	R	895330	G	895340	MG
805350	A	805360	A	805370	G	805380	A	805390	W
815350	A	815360	A	815370	W	815380	A	815390	A
825350	A	825360	G	825370	G	825380	A	825390	A
835350	G	835360	G	835370	A	835380	A	835390	WTR
845350	G	845360	G	845370	G	845380	G	845390	A
855350	A	855360	MG	855370	A	855380	G	855390	WTR
865350	G	865360	MG	865370	G	865380	A	865390	MG
875350	R	875360	G	875370	G	875380	G	875390	G
885350	G	885360	G	885370	G	885380	A	885390	A
895350	G	895360	G	895370	G	895380	G	895390	W

NUMBER OF OCCURRENCES

A	G	R	MG	W	WTR	S	**Total**
23	52	13	4	4	2	2	**100**

Arable: $$SE = \sqrt{\frac{23 \times 77}{100}} = \sqrt{17.7} = 4.2$$

Range of values at the 95% confidence level: 14.6 to 31.4%

Grassland: $$SE = \sqrt{\frac{52 \times 48}{100}} = \sqrt{25.0} = 5.0$$

Range of values at the 95% confidence level: 42.0 to 62.0%

Rough land: $$SE = \sqrt{\frac{13 \times 87}{100}} = \sqrt{11.3} = 3.4$$

Range of values at the 95% confidence level: 6.2 to 19.8%

A summary of the standard errors for each sample of 100 points is set out below:

Sample	A	G	R
1	4.7	5.0	2.9
2	4.6	5.0	2.4
3	4.6	5.0	2.7
4	4.2	5.0	3.4

The range of population values at the 95% confidence level may be summarised as follows:

Sample	A	Range	G	Range	R	Range
1	22.6–41.4	18.8	42.0–62.0	20	3.2–14.8	11.6
2	21.8–40.2	18.4	40.0–60.0	20	1.2–10.8	9.6
3	20.8–39.2	18.4	40.0–60.0	20	2.6–13.4	10.8
4	14.6–31.4	16.8	42.0–62.0	20	6.2–19.8	13.6

Analysis of results

1) From the four separate samples of 100 points it can be seen that results do not vary substantially for arable and grassland, but for rough land there are considerable differences in standard errors. In small samples a chance occurrence picked up in the sample would carry more weight than it would in a larger sample. The fourth sample of 100 points picked up more rough land and less arable than did the other three samples. It can be seen on the map that rough land is unevenly distributed over the area. This aspect of sampling will be considered again in the following section.

2) Each of the four samples in a systematic point sample in its own right. Together they may be called *interpenetrating samples* in that they are independent samples of the same 'population', using the same sampling procedure, i.e., systematic aligned point. When the four independent estimates of the population are compared there is fairly strong agreement. This is heartening to a beginner, although such agreement may not always occur.

In classwork different series of points can be assigned to different students and a comparison of the results will provide a rough check on accuracy. The total number of points sampled by the class will be very large, so that close estimates of the characteristics of the whole population will be possible. It must be stressed that each of the samples must adequately cover the whole of the area.

It has already been stated that the larger the number of units in the sample the smaller the standard error. The relationship between sample size and standard error can be demonstrated by taking samples of 200, 300 and 400 systematically selected points. In the following examples these samples have been obtained by combining the sample points of the four previous independent 100 point systematic samples. (It should be born in mind that this kind of test can be carried out with any of the three kinds of sampling unit, not only points).

Sample of 200 points

Land use	A	G	R	MG	W	WTR	S	Total
Sample	63	102	15	6	8	3	3	200
Percentage	31.5	51.0	7.5	3.0	4.0	1.5	1.5	100

Arable: $$SE = \sqrt{\frac{31.5 \times 68.5}{200}} = \sqrt{10.8} = 3.3$$

Range of values at the 95% confidence level: 24.9 to 38.1%

Grassland: $$SE = \sqrt{\frac{51.0 \times 49.0}{200}} = \sqrt{12.5} = 3.5$$

Range of values at the 95% confidence level: 44.0 to 58.0%

Rough land: $$SE = \sqrt{\frac{7.5 \times 92.5}{200}} = \sqrt{3.5} = 1.9$$

Range of values at the 95% confidence level: 3.7 to 11.3%

Sample of 300 points

Land use	A	G	R	MG	W	WTR	S	Total
Sample	93	152	23	9	13	4	6	300
Percentage	31.0	50.7	7.7	3.0	4.3	1.3	2.0	100

Arable: $$SE = \sqrt{\frac{31.0 \times 69.0}{300}} = \sqrt{7.1} = 2.7$$

Range of values at the 95% confidence level: 25.6 to 36.4%

Grassland: $SE = \sqrt{\dfrac{50.7 \times 49.3}{300}} = \sqrt{8.3} = 2.9$

Range of values at the 95% confidence level: 44.9 to 56.5%

Rough land: $SE = \sqrt{\dfrac{7.7 \times 92.3}{300}} = \sqrt{2.4} = 1.5$

Range of values at the 95% confidence level: 4.7 to 10.7%

Sample of 400 points

Land use	A	G	R	MG	W	WTR	S	Total
Sample	116	204	36	13	17	6	8	400
Percentage	29	51	9	3.25	4.25	1.5	2	100

Arable: $SE = \sqrt{\dfrac{29.0 \times 71.0}{400}} = \sqrt{5.1} = 2.3$

Range of values at the 95% confidence level: 24.4 to 33.6%

Grassland: $SE = \sqrt{\dfrac{51.0 \times 49.0}{400}} = \sqrt{6.3} = 2.5$

Range of values at the 95% confidence level: 46.0 to 56.0%

Rough land: $SE = \sqrt{\dfrac{9.0 \times 91.0}{400}} = \sqrt{2.0} = 1.4$

Range of values at the 95% confidence level: 6.2 to 11.8%

A summary of the standard errors for the systematic point samples of different sizes is set out below:

Sample	Number of points in sample	Land use		
		A	G	R
1	100	4.7	5.0	2.9
2	100	4.6	5.0	2.4
3	100	4.6	5.0	2.7
4	100	4.2	5.0	3.4
5	200	3.3	3.5	1.9
6	300	2.7	2.9	1.5
7	400	2.3	2.5	1.4

The range of the true percentage values at the 95% confidence level for the samples of different sizes may be summarised as follows:

Sample	No. of points in sample	A	Range	G	Range	R	Range
1	100	22.6–41.4	18.8	42.0–62.0	20.0	3.2–14.8	11.6
2	100	21.8–40.2	18.4	40.0–60.0	20.0	1.2–10.8	9.6
3	100	20.8–39.2	18.4	40.0–60.0	20.0	2.6–13.4	10.8
4	100	14.6–31.4	16.8	42.0–62.0	20.0	6.2–19.8	13.6
5	200	24.9–38.1	13.2	44.0–58.0	14.0	3.7–11.3	7.6
6	300	25.6–36.4	10.8	44.9–56.5	11.6	4.7–10.7	6.0
7	400	24.4–33.6	9.2	46.0–56.0	10.0	6.2–11.8	5.6

Analysis of results

1) The larger the sample the smaller the standard error and the closer the estimates are grouped around the correct value.
2) Doubling the sample size does not double the reliability, e.g. the standard error in sample 5 is not half that of sample 1. To halve the standard error and thus double the reliability a sample four times as large is necessary:

Sample	No. of points in sample	Standard error
Grassland 1	100	5.0
Grassland 7	400	2.5

The standard error is inversely proportional to the square root of the number of units in the sample. If the standard error is to be divided by 2 the number of units must be multiplied by four. If the standard error is to be divided by four the number of units must be sixteen times as great.

3) Precision increases quickly at first when the sample size is increased, then more slowly in the manner of the law of diminishing returns. There comes a point when increasing sample size gives very little extra precision.
4) To get back to reality and see what the figures mean, the purpose of the exercise was to estimate the areal extent of different types of land use within the stated area. Knowing the area is 100 km^2 we can convert the proportions into actual measurements. At the 95% confidence level our best estimate (from the sample of 400 points) for grassland is between 46 and 56% of the area, i.e. out of 100 km^2 between 46 and 56 km^2 are grassland. As the area is 100 km^2 the percentages are the same as the actual values. An area of any size could be examined but conversions from

percentages to actual values would be necessary. The kilometre grid with its subdivisions is, however, a very useful aid. If a different grid size is required this can be drawn on tracing paper and fixed over the maps, or a light table can be used with a printed grid.

As the purpose of sampling is to enable one to generalise or make a statement about the whole area, the size of the sample depends on the degree of precision required when applying the results of the sample survey to the whole. As the degree of precision depends on the sampling error, the size of error one is prepared to tolerate must be decided in advance. It may vary according to the purpose of the investigation. For example, if it is sufficient to know the true amount of grassland in the area considered above to within 20% then there is no point in taking a sample larger than a 100. This is, however, a wide range of uncertainty and not very useful. It is not necessary in practice to keep on taking samples of ever greater sizes to reach the required level of estimate for the whole area. Instead the size of sample needed for a stated size of standard error can be calculated from the data collected from an initial small sample survey, which should not be smaller than 30. Using the data from the first sample of 100 points the standard error for grassland was 5 giving a range of true values between 42 and 62% at the 95% confidence level. If we wish to reduce this range to 10% this would mean having a standard error of 2.5 instead of 5: 52 ± (2 X 2.5) gives a range of between 47 and 57%. The size of sample needed to achieve this can be found thus:

$$d = \sqrt{\frac{p \times q}{n}}$$

where d = desired value of standard error
p = percentage of land in a certain category
q = percentage of land *not* in that category
n = number of points in the sample

$$d^2 = \frac{p \times q}{n}$$

$$\therefore n = \frac{p \times q}{d^2}$$

Substituting the values obtained in the initial sample:

Arable:
$$n = \frac{32 \times 68}{(2.5)^2} = 348$$

The sample of 300 points achieved a standard error of 2.7 and the sample of 400 points one of 2.3. Therefore it follows that an intermediate sample would have a standard error of 2.5.

Grassland: $$n = \frac{52 \times 48}{(2.5)^2} = 399$$

We have already shown that this is true by our sample of 400 points giving a standard error of 2.5.

Rough land: $$n = \frac{9 \times 91}{(2.5)^2} = 131$$

Since rough land is less widely distributed than arable or grassland a smaller sample would have given a standard error of 2.5. Checking this value of $n = 131$ against the samples of 100 and 200 points, we see that samples 1, 3 and 5 confirm the result. However, sample 2 has achieved a standard error of less than 2.5 with only 100 points whereas to obtain a standard error of 2.5 for sample 4 we might expect to need a sample of well over 131 points. As mentioned previously, chance occurrences picked up in small samples tend to influence the results more than in larger samples. The uneven and less widespread distribution of rough land in this example illustrates this point. Sample 2 picked up the least roughland and sample 4 the most. These extreme values are reflected in the above results.

CHAPTER THREE

Random point sampling

The area used in the example in Chapter 2 is used here to illustrate random point sampling (1:25 000 Land Use Sheet 838 Coldstream). Sample points are selected at a series of 100 randomly chosen coordinates. This is done by using 4 digit columns from the random numbers table (or by throwing 4 dice 100 times, or 2 dice 200 times) as explained earlier. A start may be made anywhere on the table but thereafter numbers must be taken strictly in sequence as printed, either along the rows or down the columns. Some 4 digit figures may be repeated in the sequence. The second occurrence can either be ignored or the land use at that point can be recorded twice. The digits should first be listed as in Table 5 before starting to record land use. The first two digits refer to eastings on the grid and the last two to northings. The first digit of each pair refers to the kilometre grid lines and the second of each pair of digits refers to tenths of the kilometre squares. These are marked along the margins of the map and a long ruler will enable points in the non-marginal squares to be located easily. An alternative method would be to draw a suitable grid on tracing paper (or to use commercially produced transparent paper with a grid printed on it), mark on it the randomly chosen points, place it over the Land Use map and read off the type of land use. For example, the first random number on the sheet is 2017, the full grid reference for this point would be 820317. As the 8 and the 3 are constant on this part of the map they are not included in Table 5.

Arable:
$$SE = \sqrt{\frac{24 \times 76}{100}} = \sqrt{18.2} = 4.3$$

Therefore the estimated percentage of arable for the whole area at the 95% confidence level is in the range 24 ± (2 X 4.3)%, i.e. between 16.4 and 33.6%.
  Size of sample (n) required to give a standard error (d) of 2.5:

$$n = \frac{p \times q}{d^2}$$

$$= \frac{24 \times 76}{(2.5)^2} = 292$$

Grassland:
$$SE = \sqrt{\frac{54 \times 46}{100}} = \sqrt{24.8} = 5.0$$

Therefore the estimated percentage of grassland for the whole area at the 95% confidence level is in the range 54 ± (2 X 5.0)%, i.e. between 44.0 and 64.0%.

Size of sample required to give a standard error of 2.5:

$$n = \frac{54 \times 46}{(2.5)^2} = 397$$

Rough land: $\qquad$ $SE = \sqrt{\dfrac{9 \times 91}{100}} = \sqrt{8.2} = 2.9$

Therefore the estimated percentage of rough land for the whole area at the 95% confidence level is in the range $9 \pm (2 \times 2.9)\%$, i.e. between 3.2 and 14.8%.

Size of sample required to give a standard error of 2.5:

$$n = \frac{9 \times 91}{(2.5)^2} = 131$$

Table 5 Random sample of 100 points

Random number	Land use	Random number	Land use	Random number	Land use	Random number	Land use
2017	G	6728	A	4228	S	9625	G
7449	R	8586	MG	0449	A	9478	A
9470	G	4010	G	4931	W	6009	R
2215	G	9455	G	7815	R	8948	G
9329	R	1163	W	1218	G	7777	G
4504	G	6100	R	7797	A	2604	G
4491	A	5094	A	9949	A	1323	G
1623	G	6698	G	9102	G	3796	G
0450	G	6691	MG	6504	G	4283	A
3270	A	3358	G	1772	G	1218	G
0364	A	5249	G	5907	R	4016	G
6249	G	7498	A	0090	G	9399	G
6100	G	5026	G	9586	W	5430	W
8903	G	4946	G	9049	G	6189	W
0172	G	1965	G	3385	MG	1344	G
2756	A	6417	R	4979	G	4767	A
4905	G	1843	G	7448	R	9737	G
4974	G	6558	A	3725	A	6087	G
2026	G	7990	A	2243	A	3100	G
4887	G	0723	G	7796	S	0015	A
0872	A	9008	R	8746	G	1424	G
9597	W	5382	G	9862	A	6202	G
3799	G	9817	W	5731	G	2615	G
0579	A	0891	W	5837	A	1244	G
5585	A	3721	G	6342	A	4677	G

NUMBER OF OCCURRENCES

A	G	R	MG	W	WTR	S	Total
24	54	9	3	8	0	2	100

These results can be compared with the standard errors which were obtained for the same area by selecting an equal number of points systematically:

		A	G	R
Systematic samples of 100 points	1	4.7	5.0	2.9
	2	4.6	5.0	2.4
	3	4.6	5.0	2.7
	4	4.2	5.0	3.4
Random sample of 100 points		4.3	5.0	2.9

Size of sample needed to give standard error of 2.5:

	A	G	R
Systematic sample 1	348	399	131
Random sample	292	397	131

It can be seen that the results are substantially the same.

It is of interest to analyse a random sample of 50 points using the first 50 random numbers listed to compare results with the 100 points sample. (This is permissible for the random sample, but note that one cannot take the first 50 points from the systematic sample as they represent only half of the area. If a systematic sample of 50 were required a wider meshed grid should be used to cover the whole area.) The calculations are set out below:

	A	G	R	Other	Total
No. of occurrences	12	27	5	6	50
Percentage	24	54	10	12	100
Standard error	6.0	7.0	4.2		
Number in sample (n) for $d = 2.5$	292	397	144		

Arable:
$$SE = \sqrt{\frac{24 \times 76}{50}} = \sqrt{36.5} = 6.0$$

Therefore the estimated percentage of arable lies between 12.0 and 36.0% at the 95% confidence level.

Size of sample (n) required to give a standard error (d) of 2.5:

$$n = \frac{p \times q}{d^2}$$

$$= \frac{24 \times 76}{(2.5)^2} = 292$$

Grassland: $$SE = \sqrt{\frac{54 \times 46}{50}} = \sqrt{49.7} = 7.0$$

Therefore the estimated percentage of grassland lies between 40.0 and 68.0% at the 95% confidence level.

Size of sample required to give a standard error of 2.5:

$$n = \frac{54 \times 46}{(2.5)^2} = 397$$

Rough land: $$SE = \sqrt{\frac{10 \times 90}{50}} = \sqrt{18.0} = 4.2$$

Therefore the estimated percentage of rough land lies between 1.6 and 18.4% at the 95% confidence level.

Size of sample required to give a standard error of 2.5:

$$n = \frac{10 \times 90}{(2.5)^2} = 144$$

It can be seen that the limits within which one can estimate the true percentages are very much wider than for the 100 points samples as the standard error is greater with a small sample. However, the results confirm that from a sample of 50 one can predict the size of sample required for a stated standard error.

CHAPTER FOUR

Stratified point sampling

The advantages of the stratified point sampling in geographical study are considerable. The methods of sampling considered in the preceding chapters have enabled us to estimate proportions of stated land use types over the study area as a whole, but they have told us nothing about their *distribution* over the area. The most striking facts of distribution as seen on the map seem to be as follows:

1) There is a very uneven distribution of rough land. It occurs mainly in the south-east quadrant and there appears to be much less arable here. The land rises generally from the north-west to south-east, from 50 feet to 1000 feet, and slopes become steeper in the same direction. The rough land appears mainly on the steeper upland.
2) Grassland appears to be evenly distributed over upland and lowland.
3) In the upland region the main land use combination appears to be grassland/rough land, while in the lowland region the main combination is grassland/arable.

A stratified sample survey would enable us to check these statements, and add precision. It might also reveal facts of distribution not easily discernible by eye, or which have been overlooked in a visual examination of the map.

The procedure starts with a subdivision of the area into strata. The most useful approach is a division according to height/slope into 'upland/steep slope' and 'lowland/gentle slope' areas, as height and slope are clearly factors influencing land use. The line of subdivision has to be decided subjectively to some extent. At first sight the 500' contour appears to be a useful starting point. If this were followed rigidly it would leave some very steep slopes in the 'lowland' area. To avoid this the division is made as follows:

a) along the 500 ft. contour from 800 323 to 833 345;
b) along the main road south-east from 833 345 to 857 337;
c) along the foot of the steep north-west facing slope from 857 337 to 900 368.

The stratified area is shown in Figure 6.

As in the methods already described the sampling unit can be point, line or area and the method of selecting units can be random or systematic. The procedure is illustrated here by systematic point selection and the same grid is used as for the 400 point sample in Chapter 2. This time the land use for each stratum is recorded separately as shown in the table below. It is not essential to use the same grid for both areas, but it is convenient in this case as much of the work has already been done. The starting points for the grids could be fixed by random numbers.

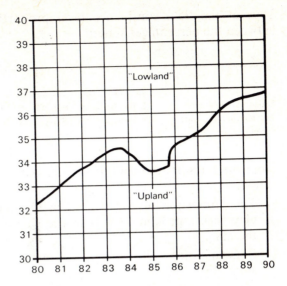

Figure 6
Stratification of the study area

UPLAND/STEEP SLOPE

	A	G	R	MG	W	WTR	S	Total
No. of occurrences	20	117	36	2	9	2	4	190
Percentage*	10.5	61.6	19.0	1.1	4.7	1.1	2.1	100.1

LOWLAND/GENTLE SLOPE

	A	G	R	MG	W	WTR	S	Total
No. of occurrences	96	87	0	11	8	4	4	210
Percentage*	45.7	41.4	0	5.2	3.8	1.9	1.9	99.9

*Correct to one decimal place

In order to find with what confidence these sample proportions can be considered representative of their areas, standard errors must be calculated. When the formula $SE = \sqrt{\frac{p \times q}{n}}$ is applied to the percentage frequency values above the results for the standard errors are:

	A	G	R
Upland/steep slope	2.2	3.5	2.8
Lowland/gentle slope	3.4	3.4	0

Thus at the 95% confidence level proportions over the whole area will be between the percentages stated:

	A	Range	G	Range	R	Range
Upland/steep slope	6.1–14.9	8.8	54.6–68.6	14.0	13.4–24.6	11.2
Lowland/gentle slope	38.9–52.5	13.6	34.6–48.2	13.6		

From these results the validity of the statements made above may be assessed as follows:

1) The proportion of rough land in the upland areas shows that it does not play such a prominent part in the land use of this area as may appear from the map.
2) Grassland does not occur as frequently in the lowland as in the upland areas. In other words the upland has a greater proportion of land under grass than the lowland.
3) In the lowland area arable is marginally more important than grassland and would predominate even more if market gardening were included in arable. In the upland area, however, grassland predominates and it is not proven from this size of sample that rough land is more dominant than arable, though it is highly likely. This is seen from the figures in the last table. If the upper possible limit for arable (14.9%) is taken and the lower possible limit of the estimation for rough land (13.4%) it could be that arable occurs more frequently than rough land.

To reduce the standard errors and thus obtain closer estimates of the true values, the sample size would have to be increased. To obtain a standard error (d) of 2.5 for grassland, for example, and at the same time automatically reduce those for other categories, a sample (n) of nearly 400 points would be necessary for each stratum:

$$n = \frac{p \times q}{d^2}$$

Upland grassland:
$$n = \frac{61.6 \times 38.4}{(2.5)^2} = 378$$

Lowland grassland:
$$n = \frac{41.4 \times 58.6}{(2.5)^2} = 388$$

Similarly, to reduce the standard error for arable in the lowland:

Lowland arable: $$n = \frac{45.7 \times 54.3}{(2.5)^2} = 397$$

These results show that to obtain a standard error of 2.5 for both upland and lowland roughly the same number of points is needed in each stratum. This emphasises the fact that it is the *number* of units in the sample which is important regardless of the relative sizes of each stratum. This point will be discussed more fully with regard to non-areal sampling.

CHAPTER FIVE

Systematic linear sampling and other methods

For the same area used in the three previous examples, the regularly spaced grid lines, either northings or eastings, may be taken as units for systematic linear sampling. Alternatively, a closer grid of evenly spaced lines can be drawn on the map if a larger sample is required.

The length occupied by each type of land use is measured along each line and recorded on the table below. Each column is totalled to obtain the total amount of the land use type on all sample lines. These values are then stated as percentages.

Northings	Length of occurrence (cm)							Total
	A	G	R	MG	W	WTR	S	
30	5.1	19.2	9.8	0.6	1.3			36
31	2.0	17.8	14.2		2.0			36
32	6.9	17.4	9.2		1.6	0.9		36
33	6.4	27.0		0.7	1.3		0.6	36
34	8.0	24.0	1.2	0.6	1.1		1.1	36
35	14.0	20.2	1.8					36
36	11.1	21.7		2.4	0.1		0.7	36
37	16.8	19.0			0.2			36
38	22.7	10.7		2.0			0.6	36
39	13.0	12.0		2.6	3.6	4.8		36
Total	106.0	189.0	36.2	8.9	11.2	5.7	3.0	360
Percentage	29.4	52.2	10.1	2.5	3.1	1.6	0.8	100

Using the percentage frequencies, the standard error can be calculated for whichever type of land use is of interest. As we have measured the land use in centimetres, the sampling unit is therefore one centimetre, thus the number of units in this sample is 360.

Arable:
$$SE = \sqrt{\frac{29.4 \times 70.6}{360}} = \sqrt{5.8} = 2.4$$

Therefore, at the 95% confidence level the estimate for the true proportion of arable is between 24.6 and 34.2%. This is practically the same estimate as that obtained by sampling 400 points systematically selected and this method of using lines is in effect the same as a point sample, the 'points' being one centimetre lengths. The coverage of the area however may be questioned as there are large tracts between the lines which cannot be included in this sampling design. Also any periodic features on the map could cause bias.

Other methods of areal sampling

The map used for the examples of areal sampling of binomial or frequency data with an overall coverage is one of the Second Land Utilisation Survey maps. These were printed from surveys made in the early 1960s. If up-to-date information is required in geographical studies some other method of obtaining it must be sought. One such method is field survey. Sampling is useful here, but sample points or lines selected on an O.S. map may be inaccessible in the field and a survey of them would be very time consuming. Some alternative methods of sampling for a field survey are considered below:

1) For practical purposes in areas well supplied with roads or tracks a route can be chosen which covers the whole area as adequately as possible. The lengths of road bordered by different types of land use can be measured by pacing or tape. The road or public track system is thus the sampling line. This will not yield an entirely unbiased sample since crops are not randomly located. For example, crops which need little attention may be sown in the more inaccessible parts of farms and consequently have no chance of selection in the sample.

2) For coverage of a wider area sampling of land use on a parish basis is possible. This involves selecting by random or systematic methods from alphabetical lists (to be found in the County Census Report) a sample number of parishes in the area being studied. The annual 4th June Agricultural Returns of the Ministry of Agriculture, Fisheries and Food can then be obtained for these sample parishes and land use records used to obtain percentage values. The acreage of each type of land use must be expressed as a percentage of the area of the parish.

3) A sample of farms in the area could be selected systematically or randomly, either from the O.S. map or from a list made from a directory, and the land use of the sample farms recorded.

II — NON-AREAL METHODS OF SAMPLING

CHAPTER SIX

Sampling procedures using interval data

Much of the data available for use in geographical studies does not originate in map form though it is amenable to mapping to reveal spatial patterns and relationships. This data can be found both in published statistics and by collecting it at first hand in the field. It is with data derived from field work that Part II of this book is concerned. The techniques illustrated in the last chapter have been concerned with the *attributes* of areas, which could be expressed in binomial form. Field surveys also produce interval data where not only is the frequency of phenomena available, but also the amount of difference or interval between one value and another. Different and more searching statistical techniques can be applied to interval data (parametric data) than to binomial data and it is essential to have some understanding of these to be able to make full and correct use of samples drawn from such data.

In areal sampling the map, or the ground covered by a grid, is the sampling frame used for the identification of individuals in the population from which the sample is selected. The first problem encountered in sampling data which does not completely cover an area, such as villages, houses, farms, is that of obtaining a record of the population; that is, a list of the number and location of all the units. Useful lists for household sampling are the registers of electors, telephone directories and street directories; for village sampling a list can be compiled from an Ordnance Survey map; for parishes from the County Census Reports. Problems immediately arise. Such lists may be defective because they are inaccurate, out of date, or incomplete and the total population may not be available. This is so when sampling a mobile population such as shoppers in a market place, or travellers in cars. Such an incomplete population is called the *sampled population*. The complete population (often not available) is the *target population*. The problem of incomplete population is common in geographical work. Strictly speaking inferences can only be made from the sample data about the sampled population. They can be extended to the target population only on the basis of substantive judgement.

An electoral list may be defective as a record of target households because it lists only the addresses of those qualified to vote. Lists compiled from the yellow pages of telephone directories, for example of farmers, bus operators or manufacturing firms, are defective because all enterprises do not subscribe to the directory. Street directories may be out of date. In practice every effort should be made to obtain a frame containing a number of units as near to the target population as possible.

The second problem is that of choosing the method to be used for selecting the sample. As in areal sampling there are no standard rules by which one can choose from the many alternatives. The selection of an appropriate procedure, whether it should be random, systematic, plain or stratified, depends on:
1) the purpose of the investigation;
2) the nature of the phenomenon being investigated;
3) feasibility in terms of practical considerations of time, available manpower and cost.

The comments made in the section on areal sampling on the advantages and disadvantages of the procedures mentioned above apply equally here. Random selection best avoids bias. On the whole systematic selection is less time consuming than random, but once the first unit is decided other units in the population do not have an equal chance of selection. Whether or not stratification is desirable depends on the purpose of the study and the nature of the data. For example, villages in an area may vary considerably in size. If size is a factor which is known or thought to affect the issue, stratification into size categories may be useful.

A third problem is that of sample size. This depends on how precisely it is desired to estimate the relevant characteristic of the population. The larger the sample the nearer one can get to the true conditions; but as has been seen in areal sampling the gain in precision does not vary directly with sample size. In general a sample should not be less than 30 units.

The ratio of the number of units in a sample to the total population is known as the *sampling fraction* (f). Therefore if, out of a total of 350 farms in an area, 70 are selected, the sampling fraction is 1/5. Where the total population is grouped into strata, use of a constant sampling fraction will ensure proportional representation in the sample. For example, a population for sampling consists of 350 textile firms of which 70 are to be selected:

	Cotton	Wool	Linen	Nylon	Total
Population	73	46	11	220	350
Sample	15	9	2	44	70

where the sampling fraction (f) = 1/5

The ratio of the population to the sample, $\frac{1}{f}$, is called the *raising factor*, (rf), and in this case is equal to five.

The problems outlined above and the basic procedures for evaluating the relationship between sample and population for discrete units and for determining what size of sample is needed for a given level of precision are considered in the following example of sampling households in a village or part of a town. They will be further illustrated by the examples in the following chapters which are based on types of field investigation commonly undertaken in geographical work.

Compiling the list

The frame from which to select the sample is a list of households in the village. This can be devised from the electoral register obtainable at a small cost from the Electoral Registration Officer usually at the local council offices. An imaginary page from such a register is shown in Table 6. The name and address of every person eligible to vote in local elections is listed and each person is numbered. The first step is to renumber the list to correspond with households to give a sampling frame. A defect here becomes apparent as more than one household may occupy a house or it may not always be clear when a house is subdivided into flats. The names may help and common sense must be used in completing the sampling frame which is shown in the last column of Table 6. In it are recorded all the individual units (households) comprising the sampled population and this is as near as we can get to the target population.

Selection procedure

Systematic selection

The starting point is decided by a random number obtained by throwing the requisite number of dice, or from a random numbers table. Thereafter households are selected at a regular interval in each direction from the first and a list is compiled. On the specimen list in Table 6, if the random number is 6 and a 20% sample is desired (i.e. $f = 1/5$), the households numbered 1, 6, 11, 16, 21, 26, etc., will form the sample. If the random number is 18 then households numbered 3, 8, 13, 18, 23, 28, 33, 38, etc. will form the sample. If a 10% sample is required (i.e. $f = 1/10$) every 10th household is listed as the sample. The problem of non-response may arise. If there is no-one at home when the investigator calls he should make three attempts at different times of day. If there is still no response this unit can be abandoned and the next house in either direction may be tried. If there is no response here after three tries this unit too must be abandoned and the sample will be one unit short. It is not permitted to keep trying next door; this is allowed only once as it would destroy the systematic element in the procedure.

Random selection

The required sample of households may be selected entirely by random numbers. If the number of households in the village is up to 100 then two twenty-sided dice, or two columns of random digits will be required. The households should be listed as they are drawn and the list will need re-arranging to save unnecessary walking from one street to another and back again. A few extra households should be selected above the number needed in the sample in case of no response. If this happens the next randomly selected one should be tried. The spares should be used strictly in order of selection.

Table 6 Sampling from the register of electors

Number	Name and address	House number	Re-numbering by household for purpose of sample
	CROFT ST.		
1	Smith A. A.	1	1
2	Jones B. B.	2	2
3	Jones C. B.	2	
4	Evans D. D.	3	3
5	Pritchard E. D.	4	4
6	Pritchard S. W.	4	
7	Jones E. W.	4	
8	Robinson T. J.	5	5
9	Robinson S. J.	6	6
10	Robinson L. T.	7	7
11	Sawyer T.	8	8
12	Sawyer S.	8	
13	Bright B. W.	9	9
14	Simpson Z. L.	9	
15	Weston R. T.	10	10
16	Bailey M. N.	11	11
	WEST ST.		
17	Parker E. T.	1	12
18	Parker D. L.	1	
19	Baker L. T.	3	13
20	Knowles C. B.	5	14
21	Edwards J. T.	7	15
22	Edwards T. F.	7	
23	Solway W. F.	9	16
	NORTH ST.		
24	Tyres S. W.	1	17
25	Porter T. A.	2A	18
26	Smart F. C.	2B	19
27	Polson H. F.	3	20
and so on, up to total number of households in settlement			

Size of sample

Before embarking upon a large time-consuming survey, full consideration must be given to the best size of sample for the investigation in hand. The size of sample necessary depends upon the size of sampling (or standard) error one is prepared to accept. It is the standard error which determines the precision with which one can estimate the parameters (inherent characteristics) of the population. Besides varying with the size of sample, the size of the standard error also varies from one sample to another according to the nature of the data. A distribution of data with a large standard deviation from the mean (that is, one in which there is a large range of occurrences on either side of the mean value) will have a larger standard error than a distribution with a small standard deviation. One cannot therefore assume that if a sample of 30 from one population (e.g. 30 households from one village) allows estimates of a given level of precision to be made for the population, a sample of 30 from another village will give equally precise estimates. The data collected from the 30 households may be differently distributed round the mean value, so that a smaller sample may give equally precise results or alternatively a larger sample may be needed. The size of sample needed can only be determined by first taking a small sample and calculating the standard error.

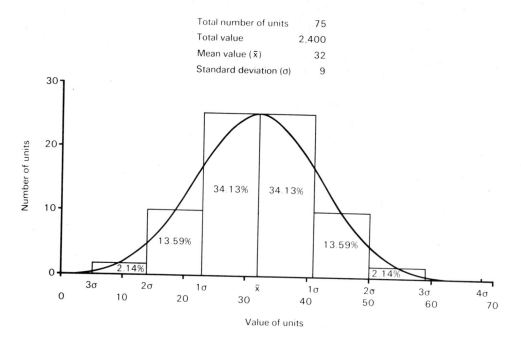

Figure 7 The normal distribution curve and histogram

SAMPLING PROCEDURES USING INTERVAL DATA

Procedures for evaluating the relationship between sample and population

To illustrate the procedures part of the data recorded from a sample of 9 from a village of 90 households is shown on Table 7. A small sample has been deliberately selected to demonstrate the procedure. The data is the result of a survey in which the investigator attempted to ascertain by means of a carefully designed questionnaire (*Appendix 5*) which towns in the area were used for goods and services not obtainable locally. The statistics refer to the number of times one of the towns was recorded by each household in the sample. This was part of a larger survey concerned with the direction of movement from the rural area to larger centres for services. The full survey will be considered in the Chapter 8.

The two parameters by which any body of interval data can be summarised are the *mean value* and the *standard deviation*. The mean value ($\bar{X}$) is the arithmetical average around which all the values are spread or dispersed. In a normal distribution values are spaced on either side of the mean as in Figure 7. The mean value ($\bar{x}$) of the data shown in column 2 of Table 7 is 7.8. The mean alone tells us little. In order to summarise the data so that we can distinguish it from any other set of data with the same mean we need an index to show by how much the values differ from the mean, or how wide is their dispersal below and above the mean. The standard deviation (σ) is the average (or standard) amount by which they differ from the mean. The formula for the standard deviation is

$$\sigma = \sqrt{\frac{\Sigma (X-\bar{X})^2}{N}} \quad \text{for the population}$$

$$\text{or} \quad s = \sqrt{\frac{\Sigma (x-\bar{x})^2}{n}} \quad \text{for the sample}$$

where X = the individual value from each unit in the population
$\bar{X}$ = the mean for the population
x = the individual value from each unit in the sample
$\bar{x}$ = the mean value for the sample
N = the number of units in the population
n = the number of units in the sample

Note that n and x refer to sample data whilst N and X refer to population data which we do not know but will estimate from the sample data. $(x - \bar{x})$ is the difference between each value and the mean of the set of values as shown in column 3. The differences are squared and summed and the average of the squared differences found $\left(\frac{\Sigma(x-\bar{x})^2}{n}\right)$. This value (20.6) is called the variance. The square root of it (4.5) is the *sample standard deviation* (*s*).

These two parameters, the mean and the standard deviation neatly summarise the data in a normal distribution. In such a distribution (*Figure 7*):

1) Approximately 68% of the values occur within one standard deviation of the mean, 34% of them are larger than the mean and 34% smaller than the mean.
2) 95% of the values occur within two standard deviations of the mean.
3) 99.7% occur within three standard deviations of the mean.
4) 99.99% occur within four standard deviations of the mean.

A small sample may not correspond to the normal distribution. The larger the sample the more nearly will it correspond to the population distribution around the mean.

The sample standard deviation on Table 7 is 4.5. The range within one standard deviation of the mean is from 7.8−4.5 to 7.8 + 4.5, i.e. 3.3 to 12.3 and would be expected to include 68% of the values. Similarly the range 7.8−(2 × 4.5) to 7.8 + (2 × 4.5), i.e. 0 to 16.8, would include at least 95% of the values. This range does in fact include all the sample values in this example. It can be seen that a larger standard deviation would indicate a wider spread of values around the mean.

The task is now to use these sample parameters to estimate the parameters of the population, $\bar{X}$ and σ.

Table 7 Data from household survey

Household	Number of mentions*	$(x-\bar{x})$	$(x-\bar{x})^2$
1	15	7.2	51.8
2	10	2.2	4.8
3	4	−3.8	14.4
4	2	−5.8	33.6
5	14	6.2	38.4
6	4	−3.8	14.4
7	11	3.2	10.2
8	6	−1.8	3.2
9	4	−3.8	14.4
Mean	7.8	Total	185.2

*of the main centre used

$$\therefore s = \sqrt{\frac{\Sigma (x-\bar{x})^2}{n}}$$

$$= \sqrt{\frac{185.2}{9}} = 4.5$$

SAMPLING PROCEDURES USING INTERVAL DATA

The Mean ($\bar{X}$)

If the sample is large enough (at least 30 units) the sample mean plus or minus two standard or sampling errors will include 95% of the probable values of the true or population mean. We can thus say at the 95% confidence level that the true mean lies within two standard errors of the sample mean. The smaller the standard error therefore the more accurate is the sample mean as an estimate of the true mean.

Standard deviation (σ)

The spread of values round the mean in the population is almost certain to be greater than the sample spread, especially if only a small sample has been taken. The sample standard deviation(s) will therefore be less than the population standard deviation (σ). The correction $\sqrt{\frac{n}{n-1}}$ is thus applied to the sample value to give a best estimate ($\hat{\sigma}$) of the true value of the standard deviation:

$$\hat{\sigma} = s \times \sqrt{\frac{n}{n-1}}$$

$$= \sqrt{\frac{\Sigma(x-\bar{x})^2}{n}} \times \sqrt{\frac{n}{n-1}}$$

$$= \sqrt{\frac{\Sigma(x-\bar{x})^2}{n-1}}$$

Therefore in the above example:

$$\hat{\sigma} = \sqrt{\frac{185.2}{8}} = 4.8$$

It is clear that n is all important, as it is the number of units in a sample which determines the precision with which we can estimate the parameters of the population. Whether this number represents 9% or 19% of the population $\frac{n}{n-1}$ becomes nearer to unity the larger the value of n and therefore s becomes closer to $\hat{\sigma}$.

Standard error of the mean (SEM)

The value of the true mean will range within certain limits from the sample mean. The value which determines these limits is known as the standard error of the mean (SEM). The formula for the standard error of the mean is

$$\text{SEM} = \frac{\hat{\sigma}}{\sqrt{n}}$$

Therefore in the above example:

$$\text{SEM} = \frac{4.8}{\sqrt{9}} = 1.6$$

Summarising, we now have:
- a) the sample mean, $\bar{x} = 7.8$;
- b) the best estimate of the standard deviation, $\hat{\sigma} = 4.8$;
- c) the standard error of the mean, SEM = 1.6.

If the sample were a large one it would be possible to state from the normal distribution at the 95% confidence level that

$$\bar{X} = \bar{x} \pm 2 \times \text{SEM}$$

However, the sample in the above example is small (a small sample is one with less than 30 units) and it is not safe to assume that the true value of the mean will lie within two standard errors of the sample mean with this degree of confidence. Thus instead of the normal distribution the Student's t distribution must be used and the appropriate value of t inserted in the above equation in place of the number 2. The index t represents the relationship between the difference of the sample and population means and the standard error of the mean:

$$t = \frac{\text{difference of means}}{\text{standard error of the mean}} = \frac{\bar{X} - \bar{x}}{\left(\dfrac{\hat{\sigma}}{\sqrt{n}}\right)}$$

To refer to Student's t graph or tables (see *Appendix 4*) it is necessary to know the 'degrees of freedom'. This is $n - 1$ and it refers to the number of values which can be assigned arbitrarily without changing the value of the mean. That is, in a sample of 9, once 8 values are established the 9th value is automatically fixed. If a 95% confidence level is used t is read off opposite 8 degrees of freedom in the 5% column, giving $t = 2.3$. It will be noticed on this table that with more than 30 degrees of freedom (i.e. with a sample of more than 31 units) the value of t remains at almost 2. Therefore the true mean in this small sample is:

$$\bar{X} = \bar{x} \pm t \times \text{SEM}$$
$$\therefore \bar{X} = 7.8 \pm 2.3 \times 1.6$$
$$= 7.8 \pm 3.7$$

i.e. at the 95% confidence level $\bar{X}$ lies between 4.1 and 11.5. From the small sample survey all we can say for the population of 90 households in the village

SAMPLING PROCEDURES USING INTERVAL DATA

is that the average number of mentions per household of town A is between 4 and 12 to the nearest whole number. We can say this with a confidence level of 95%. Therefore the total number of mentions from the population is between 360 and 1080. This is not very useful but it is impossible to give a closer estimate at the 95% confidence level from so small a sample.

Calculating the size of sample necessary for closer estimates

Choosing the right size of sample for the purpose in hand is one of the arts of sampling. Having found the standard error of the mean for a small sample it is possible to use it to determine the sample size needed to give any required standard error.

The range within which the population mean lies is 4 to 12, i.e. a range of 8. If we wish to be able to state the true mean within a range of only 4, i.e. 2 on either side of the mean, we need a standard error of 1. The size of sample which would give this can be found as follows:

$$d = \frac{\hat{\sigma}}{\sqrt{n}}$$

where d = desired value of SEM

$$\therefore \sqrt{n} = \frac{\hat{\sigma}}{d}$$

$$n = \left(\frac{\hat{\sigma}}{d}\right)^2$$

Substituting the values for $\hat{\sigma}$ and d given above:

$$n = \left(\frac{4.8}{1}\right)^2 = 23$$

This is in accordance with the statement that the standard error is inversely proportional to the square root of the number of units in the sample, i.e. to divide the standard error by two, the number in the sample must be multiplied by the square of two. A sample four times the size of the original one would produce a standard error of 0.8, i.e. half the original value. For a sample of 36:

$$\text{SEM} = \frac{\hat{\sigma}}{\sqrt{n}}$$

$$= \frac{4.8}{\sqrt{36}} = 0.8$$

One other matter must be mentioned to complete the procedures. The *standard error of the standard deviation* can be calculated within confidence limits though this is not as important as the standard error of the mean. For practical purposes it is usually sufficient to use the best estimate of the standard deviation. If the outside limits of the possible dispersion range of population values around the mean are important to the study these can be calculated from the standard error of the standard deviation. It will be apparent however that where a small sample has been taken and consequently the population mean has a large range of possibility, discussion of the possible range of standard deviation values as well would become very complex. If a large sample has been taken however, and a great degree of accuracy is required in estimations, the following formula gives the degree of uncertainty inherent in the estimate of the true standard deviation. The formula for the standard error of the standard deviation (SE of s) is:

$$\text{SE of } s = \frac{\hat{\sigma}}{\sqrt{2n}}$$

Using the values of $\hat{\sigma}$ and n for the sample of 9 households:

$$\text{SE of } s = \frac{4.8}{\sqrt{18}} = 1.1$$

The true standard deviation will be in the range $s \pm (2 \times 1.1)$ at the 95% confidence level, i.e. 2.3 to 6.7, where $s = 4.5$.

It is apparent from the explanation given that the difference between sample and true parameters are affected by three sets of circumstances:

1) *The size of the standard deviation.* This is inherent in the data and cannot be altered. One needs a larger sample when data is widely dispersed no matter what the size of the population. Thus if two villages of the same size were being sampled for comparative purposes, one with values widely dispersed around the mean and the other with values clustered close to the mean, the former would need a larger sample than the latter to achieve comparable standard errors. A survey using a uniform sampling fraction, i.e. giving the villages equal representation in the sample, would not be satisfactory.

2) *The confidence level selected.* If one wishes to state population parameters at the 99% confidence level one must accept wider limits for the mean, or select a large sample.

3) *Sample size.* The investigator has a choice of the size of the standard error and the confidence level. The choice depends on the purpose of the analysis and the degree of precision required.

The art of sampling lies in choosing the minimum size of sample which will give an answer with a degree of confidence adequate for the purpose. If a certain degree of precision is required then a certain size of sample is necessary. An adequate answer cannot be obtained from an inadequate size of sample. There is only one way to find the *exact* characteristics of a population and that is to study the whole, but from a sample one can estimate to within *stated limits*. The larger the sample, the smaller the limits and the nearer to the true conditions. Examples in the following chapters will illustrate this.

CHAPTER SEVEN

Field sampling from an unknown population total

To demonstrate this sampling technique, a questionnaire type survey of shoppers in Lincoln town centre has been used in the following example. The survey is part of an investigation into the rôle of the town as a central place.

Method

The key question to be asked is where do the selected units (shoppers) come from. Further information may be gathered as to how often the shopper visits the centre, whether for low or high order goods and which other centres are used. The first step in organisation is to consider the most useful form in which to pose such questions. If the shopper lives in the town the street should be given; for shoppers from the surrounding area the name of the settlement is required for precise mapping of the results.

Selecting the sample

In selection of sample units it is difficult to apply strict random or systematic methods to a moving population. The number of interviews possible in the time and with the manpower available should be assessed and the best location chosen for the job. If possible interviewers should work in pairs and units should be selected as far as possible at a regular interval of every fifth or every tenth person to enter the store or market. Alternatively, as many as possible in a given time may be the only practicable means of coverage. A market day, when extra bus services run from the surrounding area is a suitable day to choose.

    A similar technique can be applied in a survey of leisure seekers at a resort or in a National Park, where a car or bus park is a suitable place at which to carry out the enquiry.

Tabulation and processing

Figure 8 shows the mapped results of part of the data collected in a shoppers' survey in Lincoln on a market day. 400 shoppers were interviewed. Of these 284 (71%) came from Lincoln and its contiguous built up area. 116 (29%) came in from outlying settlements. It is of interest to look at the distance from which people regularly come in to shop. On the map circles have been drawn at intervals of 8 km from Lincoln city centre. Table 8 shows the frequency of shoppers from different distances. To assess to what extent these findings can be considered valid for the population it is necessary to calculate standard

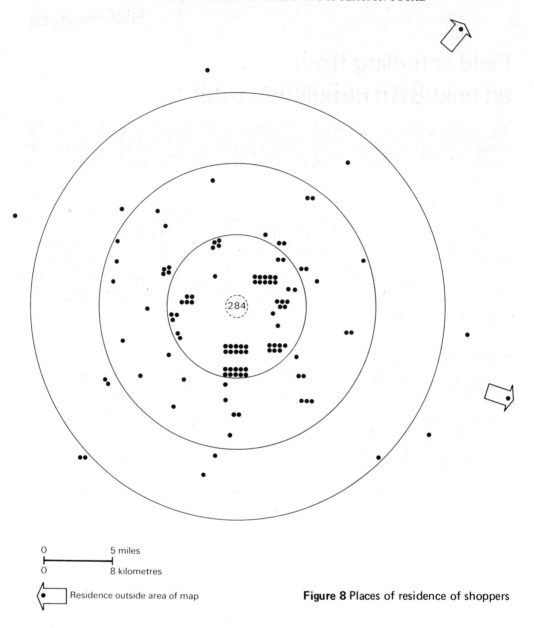

Figure 8 Places of residence of shoppers

errors by the method used for frequency data. The calculations are shown below Table 8. We can now say at the 95% confidence level that between 66 and 76% are shoppers from the town, while between 24 and 34% come in from the surrounding area. In other words this survey shows that between a quarter and a third of Lincoln's trade is from outside the city area. The standard error of the 16% in the sample who came from within 8 km is 1.8, showing that from

12.2 to 19.4% come from within this distance. 6.5 to 12.5% come from 8 to 16 km and 1.8 to 5.8% from further afield. Further work on the comparison of the sample values and the central place function of Lincoln will be considered in Part III.

Table 8 Lincolnshire shopping survey: Place of residence of shoppers

	Lincoln	up to 8 km	8 to 16 km	over 16 km	Total
Number of shoppers	284	63	38	15	400
Percentage*	71.0	15.8	9.5	3.8	100.1

*Percentages are correct to one decimal place

Lincoln:
$$SE = \sqrt{\frac{p \times q}{n}}$$

where p = percentage from Lincoln
q = percentage not from Lincoln
n = total number in sample

$$SE = \sqrt{\frac{71 \times 29}{400}} = 2.3$$

The population percentage coming from Lincoln lies between 71 ± (2 × 2.3) at the 95% confidence level, i.e. between 66.4 and 75.6%.

Up to 8 kilometres:
$$SE = \sqrt{\frac{15.8 \times 84.2}{400}} = 1.8$$

The population percentage from within a radius of up to 8 km is between 12.2% and 19.4%.

8–16 kilometres:
$$SE = \sqrt{\frac{9.5 \times 90.5}{400}} = 1.5$$

The population percentage coming from between 8 and 16 km of Lincoln is in the range 6.5 to 12.5%.

Over 16 kilometres:
$$SE = \sqrt{\frac{3.8 \times 96.2}{400}} = 1.0$$

The population percentage coming from further than 16 km away from Lincoln is between 1.8 and 5.8%.

CHAPTER EIGHT

Household sampling with uniform sampling fraction

The aim of this example is to determine the direction and volume of movement from a rural area to nearby larger centres for goods and services. The area concerned is that on either side of the A15 north of Lincoln, containing twenty villages. For this study a village is defined as a settlement with a post office. The location of the village is shown on Figure 9 and the number of households in each village on Table 9.

Method

A survey questionnaire is used on a sample of households in each village (*Appendix 5*). The questionnaire is in six parts:

Part 1 Centres used for the purchase of 5 types of low order goods.

Part 2 Centres used for the purchase of 6 types of high order goods.

Part 3 Centres used for 10 types of service.

Part 4 Centres used for 4 types of entertainment.

Part 5 Centres used for work and school.

Part 6 Centres of origin of newspapers taken.

When the results are processed aspects can be considered separately or the total number of mentions of different centres can be considered and the findings tabulated for mapping.

Selecting the sample

There are three possible alternatives for the sampling frame:
1) a list of households compiled from Electoral Registers;
2) a list compiled from a street directory;
3) the street plan on the ground or a 1:2500 map updated by field work.

Alternative 3) is only suitable for a systematic sample, the selection of units being made in the field. Alternative 1) will be found to be most easily available and has been used in this example.

   Another consideration is that of the size of sample to be taken and the method of selection. The manpower and time available is sufficient to cope with approximately 400 house enquiries. The possible alternative methods for selection from each village are as follows:

1) To use the same number of households from each village; in this case 20 would be possible for the manpower available.

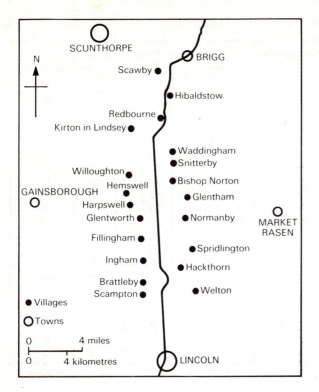

Figure 9
Location of villages in the survey

2) To use a number proportional to the population (number of households) in each village. A uniform fraction of 1/10 would give approximately 400 in all. The actual number would vary from 70 in the largest village to 3 in the smallest.
3) To use a variable sampling fraction, selecting a larger proportion of the population from smaller villages.

Alternative 2) is used here to show the limitations of this method. The importance of the *number* in the sample and the advantages of method 1) will become apparent during the processing.

A final decision has to be taken as to whether the sample will be random, systematic or random stratified. The data is already stratified on the basis of location in that the units (households) are grouped into 20 different villages. For the purpose in hand an individual sample is required from each village and each will be treated as an individual population. Uses of population stratification will be considered later as more complicated calculations are necessary.

In field sampling it is essential to consider thoroughly the probable merits or demerits of the possible alternatives before starting, as it may be impossible to go back to take a larger sample if the data proves unsatisfactory for the level of confidence required. It is not always possible to take an initial small sample

Table 9 Lincolnshire household survey: Summary of sample data

Village	Number of households	Number in sample	NUMBER OF MENTIONS						Dominant centre
			L	B	S	G	MR	Others	
1 Kirton Lindsey	702	70	54	90	413	54		46	S
2 Scawby	513	43	15	445	332	1	1	33	B
3 Scampton	377	38	92	1		4		16	L
4 Hibaldstow	349	35	23	284	281	4		33	B/S
5 Hemswell	349	35	51	7	25	148		15	G
6 Welton	302	30	443					29	L
7 Ingham	198	20	277		2	14	1	20	L
8 Waddington	173	17	17	111	101	12		12	B/S
9 Willoughton	146	14	47	10	46	90		14	G
10 Glentworth	122	12	88	3	15	72		15	L/G
11 Glentham	100	10	75	1	5	2	45	20	L
12 Redbourne	99	10	9	71	37	3	1	6	B
13 Bishop Norton	90	9	70	20	26	11	9	5	L
14 Snitterby	74	7	27	45	5	2	5	14	B
15 Hackthorn	69	7	110				1		L
16 Fillingham	65	7	79	1	2	23	1	2	L
17 Harpswell	53	5	38	3		48		5	G
18 Spridlington	52	5	44		4	1	8	7	L
19 Brattleby	29	3	44					2	L
20 Normanby	26	3	41		1	1	3	2	L

B Brigg   G Gainsborough   L Lincoln   MR Market Rasen   S Scunthorpe

to find out what size is necessary for making statements about the population within acceptable limits. When sampling from data contained in published lists this is easy to do. The purpose of the exercise must be kept in mind and all factors relevant to it considered.

A problem encountered in a survey of this kind is the best time of day to undertake it. On weekdays there is likely to be a good deal of non-response due to the members of the household being out in working hours. A systematic sample would not be suitable unless it was possible to return in the evening. A random sample on the other hand could give a biased view by including an undue proportion of retired people if many numbers had to be discarded and replaced by the extras until the required number were obtained.

It is now obvious that there can be no hard and fast rules as to which type and size of sample are best. The investigator must weigh up all relevant factors and consider all possible problems before deciding.

Tabulation and processing

A large amount of data was collected in this survey and it can be used in a variety of ways. If a very detailed study of movements for different types of services is required, the answers to each of the six sections of the questionnaire should be tabulated separately. If it is hoped to produce a map showing a picture of overall movement the total number of mentions for each of the five centres should be tabulated for each household sampled. If movement to centres outside the area is of interest these should be tabulated separately too.

Much inter-village movement was discovered and recorded and this could form the basis of another study. It can be seen that one sample survey, if properly planned, conducted and processed, can reveal sufficient information for a detailed study of many aspects of an area.

The data tabulated in Table 9 summarises the overall movement. The total number of mentions of each of the five centres has been extracted from the household sample sheets and the mean and the standard error of the mean calculated. Table 10 shows this for two of the villages and a summary of the results for 11 of the villages is shown on Table 11. The main point to emerge concerning the usefulness of the data from this type of sampling (uniform fraction) is that the standard error varies considerably from one village to another. The small samples, for example those from Harpswell and Spridlington, show a larger standard error than the larger samples such as those from Scawby, Welton and Ingham. Figure 10 shows the relation between size of samples and standard errors. It will be noticed that the lower standard deviations of Glentham and Fillingham have resulted in lower standard errors than is the case with other samples of similar size. The standard error which would result from a sample of 20 from each village has been calculated and is approximately one. A sample of this size therefore would enable the population mean to be stated within similar limits for all villages.

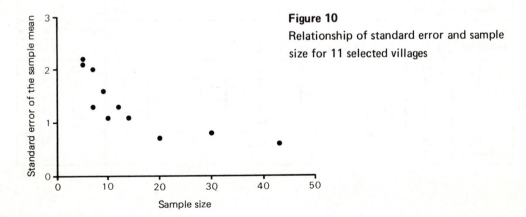

Figure 10

Relationship of standard error and sample size for 11 selected villages

Table 10 Lincolnshire household survey: Data for Welton and Harpswell

WELTON Population: 302 households; 30 in sample

Household	NUMBER OF MENTIONS						Lincoln data	
	Lincoln	Brigg	Scunthorpe	Gainsborough	Market Rasen	Others	$(x-\bar{x})$	$(x-\bar{x})^2$
1	17						+2.2	4.84
2	14					3	−0.8	0.64
3	15						+0.2	0.04
4	21						+6.2	38.44
5	12						−2.8	7.84
6	18						+3.2	10.24
7	13						−1.8	3.24
8	7					5	−7.8	60.84
9	18						+3.2	10.24
10	21						+6.2	38.44
11	6					6	−8.8	77.44
12	18						+3.2	10.24
13	15						+0.2	0.04
14	8					3	−6.8	46.24
15	18						+3.2	10.24
16	15						+0.2	0.04
17	5					2	−9.8	96.04
18	14						−0.8	0.64
19	12						−2.8	7.84
20	8					3	−6.8	46.24
21	16					1	+1.2	1.44
22	11					1	−3.8	14.44
23	18						+3.2	10.24
24	17					1	+2.2	4.84
25	26						+11.2	125.44
26	15					2	+0.2	0.04
27	17						+2.2	4.84
28	14					1	−0.8	0.64
29	18						+3.2	10.24
30	16					1	+1.2	1.44
Total	443					29		643.40
Mean	14.8					1.0		

HARPSWELL Population: 53 households; 5 in sample

							Gainsborough data	
							$(x-\bar{x})$	$(x-\bar{x})^2$
1	10			6			−3.6	12.96
2	7			7		1	−2.6	6.76
3	13	2		6		1	−3.6	12.96
4	3	1		12		1	+2.4	5.76
5	5			17		2	+7.4	54.76
Total	38	3		48		5		93.20
Mean	7.6			9.6				

Welton:
$$\hat{\sigma} = \sqrt{\frac{\Sigma (x-\bar{x})^2}{n-1}}$$
$$= \sqrt{\frac{643.4}{29}} = \sqrt{22.2} = 4.7$$

$$\text{SEM} = \frac{\hat{\sigma}}{\sqrt{n}}$$

$$= \frac{4.7}{\sqrt{30}} = 0.8$$

At the 95% confidence level:

$$\bar{X} = 14.8 \pm (2 \times 0.8)$$
$$= 14.8 \pm 1.6$$
$$= 13.2 \text{ to } 16.4$$

i.e. the mean value for the whole population lies between 13.2 and 16.4. A sample of 20 would have given a standard error of 1.1.

Harpswell:
$$\hat{\sigma} = \sqrt{\frac{93.2}{4}}$$
$$= \sqrt{23.3} = 4.8$$

$$\text{SEM} = \frac{4.8}{\sqrt{5}} = 2.2$$

At the 95% confidence level:

$$\bar{X} = 9.6 \pm (t \times 2.2)$$

(t being used because this is a small sample). For 4 degrees of freedom at the 95% confidence level:

$$t = 2.8$$
$$\therefore \bar{X} = 9.6 \pm 6.2$$
$$= 3.4 \text{ to } 15.8$$

i.e. the mean value for the whole population lies between 3.4 and 15.8. A sample of 20 would give a standard error of 1.1.

Table 11 Lincolnshire household survey: Summary of calculations for selected villages

Village	No. in sample	Dominant centre	Sample mean	Standard deviation	SE of sample mean	Limits of estimated population mean at 95% confidence level	SE which would be obtained from sample of 20
Scawby	40	B	10.3	4.0	0.6	9.1 to 11.5	0.9
Welton	30	L	14.8	4.7	0.8	13.2 to 16.4	1.1
Ingham	20	L	13.9	3.1	0.7	12.4 to 15.4	0.7
Willoughton	14	G	6.4	4.3	1.1	4.0 to 8.8	1.0
Glentworth	12	L	7.3	4.6	1.3	4.4 to 10.2	1.0
Glentham	10	L	7.5	3.6	1.1	5.0 to 10.0	0.8
Bishop Norton	9	L	7.8	4.8	1.6	4.1 to 11.5	1.1
Hackthorn	7	L	15.7	5.2	2.0	10.7 to 20.7	1.2
Fillingham	7	L	11.3	3.5	1.3	8.1 to 14.5	0.8
Harpswell	5	G	9.6	4.8	2.2	3.4 to 15.8	1.1
Spridlington	5	L	8.8	4.7	2.1	2.9 to 14.7	1.0

It can be seen that in a study such as this the use of the available manpower and time to take a sample of 20 from each village would have given more useful overall results.

CHAPTER NINE

Household sampling with variable sampling fraction

In the following example, using the same questionnaire as in the previous chapter, a similar study of villages in part of Derbyshire is considered. This differs from the Lincolnshire example in that here a random sample of the same number of households (30) was selected from each village, regardless of the population of the villages. For one village (Elton) the full household data is shown on Table 12 and the credentials of the sample established. The data collected from other villages is summarised on Table 13. For Elton the standard error of the mean for the number of mentions of Matlock is 0.5 giving a population mean of between 5.7 and 7.7 estimated at the 95% confidence level. The sample from other villages yielded similar standard errors from 0.5 to 0.7. These have been calculated for the town with the largest number of mentions in each case.

Mapping from the sample data

Cartographic problems arise when an attempt is made to map the results of the survey. The object of the exercise as stated on page 00 was to estimate within acceptable limits the direction and volume of movement from a rural area to nearby large centres for goods and services. The most appropriate method is by flow lines of a width proportional to the volume of movement. For each centre an estimated number for the volume can be calculated by multiplying the sample mean by the number in the population; for Elton to Matlock this is 6.7 $\times$ 151 = 1012. The possible error can be stated by using the limits of the population mean. For Elton the standard error is 0.5 which indicates that the population mean lies between 5.7 and 7.7 at the 95% confidence level. If the lower estimate is taken the volume would be 5.7 $\times$ 151 = 861. If the higher estimate is taken the volume would be 7.7 $\times$ 151 = 1163.

An alternative method of mapping the data would be to use pie graphs, by which method each village would be represented by a circle with radius proportional to the square root of its population. Each circle can be divided into segments proportional to the number of mentions of the centres used. The method does not give the impression of movement as well as a flow line map, but it may be more practicable if villages are near together.

One further very important consideration is necessary before using the sample data for mapping. The differences revealed between the villages must be proved to be significant or real differences. They could be chance occurrences. This matter will be dealt with in Chapter 12 and the sample results will be mapped only if they are valid.

Table 12 Derbyshire household survey: Data for Elton

Population: 151 households; 30 in sample

Household	NUMBER OF MENTIONS					Matlock data	
	Matlock	Bakewell	Chesterfield	Sheffield	Derby	$(x-\bar{x})$	$(x-\bar{x})^2$
1	8	3	3			1.3	1.69
2	9	6		3		2.3	5.29
3	2	2			1	4.7	22.09
4	7	2	2			0.3	0.09
5	6	3	6			0.7	0.49
6	6	3	4			0.7	0.49
7	12	4	3			5.3	28.09
8	7	4		1	3	0.3	0.09
9	4	3		5	1	2.7	7.29
10	2	13				4.7	22.09
11	10	3	3	1		3.3	10.89
12	10	5	4			3.3	10.89
13	5	6	2			1.7	2.89
14	5		2			1.7	2.89
15	4		2			2.7	7.29
16	3	6	3			3.7	13.69
17	3	5	2	2		3.7	13.69
18	9	5	1		3	2.3	5.29
19	7	1	1			0.3	0.09
20	13	2				6.3	39.69
21	9	1	1	2	1	2.3	5.29
22	7					0.3	0.09
23	8		1		1	1.3	1.69
24	5	2				1.7	2.89
25	7	1	1			0.3	0.09
26	5	4		1		1.7	2.89
27	7	2		1		0.3	0.09
28	9	3		1		2.3	5.29
29	6	3				0.7	0.49
30	7	4	1			0.3	0.09
Total	202	96	42	17	10		213.90
Mean	6.7						

$$\hat{\sigma} = \sqrt{\frac{213.9}{29}}$$

$$= \sqrt{7.4} = 2.7$$

$$\text{SEM} = \frac{2.7}{\sqrt{30}} = 0.5$$

Table 13 Derbyshire household survey: Data for Matlock area

Village	Number of households	Number in sample	Number of mentions					
			Matlock	Bakewell	Chesterfield	Sheffield	Derby	Buxton
Elton	151	30	202*	96	42	17	10	
Beeley	77	30	148*	100	117	60	21	
Bonsall	287	30	366*	2	37	3	31	
Tansley	224	30	510*	4	58	25	29	
Monyash	114	30	4	242*	25	29	5	224

Village	Mean number of mentions per household ($\bar{x}$)						SEM for dominant centre
	Matlock	Bakewell	Chesterfield	Sheffield	Derby	Buxton	
Elton	6.7*	3.2	1.4	0.6	0.3		0.5
Beeley	4.9*	3.3	3.9	2.0	0.7		0.5
Bonsall	12.2*	0.1	1.2	0.1	1.0		0.7
Tansley	17.0*	0.1	1.9	0.8	1.0		0.5
Monyash	0.1	8.1*	0.8	1.0	0.2	7.5	0.6

Village	Number of mentions from whole population estimated from sample data ($\bar{x}$ × no. of households)						Range of ($\bar{X}$ × no. of households) at 95% confidence level
	Matlock	Bakewell	Chesterfield	Sheffield	Derby	Buxton	
Elton	1012*	483	211	91	45		861–1163
Beeley	377*	254	300	154	54		300– 454
Bonsall	3501*	29	344	29	287		3099–3903
Tansley	3807*	22	426	179	224		3584–4032
Monyash	11	923*	91	114	23	855	1060– 786

*Data for dominant centre

CHAPTER TEN

Sampling in a study of farm economy

In a study of farming one or two farms are often selected purposively to be visited by a class of students. With senior students working in pairs and well briefed, a better coverage method would be to make a random selection of farms in the area. In most areas there is no 'typical farm' and this method, which has been used in the example below, brings out variety in farm size and farm practice.

Selecting the sample

The first problem is obtaining a sampling frame, that is, a list of farms in the area concerned, from which to select the sample. If no exclusive list is obtainable the yellow pages of the relevant telephone directories can be used satisfactorily. These would provide the sampled population, though it must not be regarded as the target population as all farms may not subscribe. Farms should be listed and numbered in alphabetical order and the selection made by use of random numbers. In the following example using a raising factor of 4, 38 farms in part of Derbyshire were selected from a list of 152 compiled in this way. For the questionnaire used to gather the information see Appendix 6.

Tabulation and processing

Figure 11 shows the farms plotted in rank order of size and Table 14 gives a summary of selected data obtained from the questionnaires. It can be seen immediately that a great deal of data concerning farm size and farm practice has been gathered and it is necessary to estimate the confidence with which the facts found in the sample are applicable to the population.

SIZE

Figure 11 shows that farms vary from 21 to over 1000 acres, so that to calculate the mean and standard deviation for farm size would seem to be of little use. As the range is so great the standard deviation, and thus the standard error of the mean, would be very large. Therefore the limits of the estimates for the mean and for the range of farm size in the population would be so wide at an acceptable confidence level that they would be meaningless. On the graph the farms can be seen to fall into 3 groups and more precise information may be obtained about size of farms in the population by treating size as an attribute of the farm. The formula for frequency data (*Chapter 2*) can be applied to find out the likely proportion of the population of farms in each of the size categories.

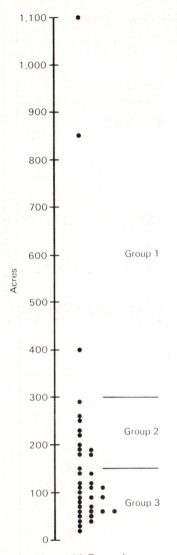

Figure 11 Farm size

Group 1 Over 300 acres
Out of a sample of 38, 3 are in this group.

$$p = \frac{3}{38} \times 100 = 7.9\% \therefore q = 92.1\%$$

$$SE = \sqrt{\frac{p \times q}{n}}$$

$$SE = \sqrt{\frac{7.9 \times 92.1}{38}} = \sqrt{19.1} = 4.4$$

At the 95% confidence level the proportion of farms of over 300 acres is 7.9 ± (2 × 4.4), i.e. between 0 and 16.7%. This means that the chances are 95 in 100 that less than about 17% of the farms in the area are over 300 acres.

Group 2 151–300 acres
There are 10 farms in this group in the sample.

$$p = \frac{10}{38} \times 100 = 26.3\% \therefore q = 73.7\%$$

$$SE = \sqrt{\frac{26.3 \times 73.7}{38}} = \sqrt{51.0} = 7.1$$

At the 95% confidence level the proportion of farms of between 151 and 300 acres is 26.3 ± (2 × 7.1), i.e. between 12.1 and 40.5%. The chances are therefore that in 95 cases out of 100 between approximately 12 and 40% of the farms in the area are between 151 and 300 acres.

Group 3 Farms of 150 acres or less
The remaining 25 farms in the sample are in this group.

$$p = \frac{25}{38} \times 100 = 65.8\% \therefore q = 34.2\%$$

$$SE = \sqrt{\frac{65.8 \times 34.2}{38}} = \sqrt{59.2} = 7.7$$

At the 95% confidence level the proportion of farms in the area of 150 acres or less is between 65.8 ± (2 × 7.7), i.e. between 50.4 and 81.2%. In other words the chances are 95 in 100 that more than half the farms in the area are 150 acres or less and there could be as many as 81%.

Table 14 Farm survey: Summary of data

GROUP 1 FARMS OVER 300 ACRES

Farm number	Acreage	ARABLE ACREAGE						GRASS ACREAGE		LIVESTOCK NUMBERS				LABOUR	
		Wheat	Barley	Oats	Rye	Potatoes	Kale	Temp.	Perm.	Dairy cattle	Other cattle	Sheep	Poultry	Full time	Part time
1	1100	60	220					170	820	240	163			7	
2*	850	7						328	490	90	84	350	40	5	1
3	402		19		7		2		55	32	141	108		4	

*Farm 2 had 174 acres of rough land not in agricultural use

GROUP 2 FARMS BETWEEN 151 AND 300 ACRES

Farm number	Acreage	Wheat	Barley	Oats	Rye	Potatoes	Kale	Temp.	Perm.	Dairy cattle	Other cattle	Sheep	Poultry	Full time	Part time
4	286								286	60	68	48	10	3	
5	260		25						260	20	160	150		1	1
6	250								225	42				2300	3	
7	230								230	56	32				
8	220								220	140		23	80	2	
9	202								202	50	70	80	24	2	
10	190							130	60		60	48			
11	188								180	35	47	30		1	
12	182							90	92	98		226		2	1
13	180								180	55		40	150	2	1

GROUP 3 FARMS OF 150 ACRES OR LESS

Farm number	Acreage	Wheat	Barley	Oats	Rye	Potatoes	Kale	Temp.	Perm.	Dairy cattle	Other cattle	Sheep	Poultry	Full time	Part time
14	150		12					31	99	50	53		50	3	
15	136		3					118		38	10		60	2	1
16	135		10			3	5	60	60	130	5		80	3	
17	123					7	3		83	18	13	24	20	2	
18	120		20				4		50	70	12	15		1	1
19	113							50	50	100			100	2	
20	108							33	80	60	60			2	
21	105		15			4		83	20	40	20		30	1	1
22	100					2			81	55	26		150	2	
23	90								100	36			12	1	1
24	90		20					48	42	45	6			1	1
25	87							60		25	12		20	1	
26	76								87	52	20		12	1	
27	65								76	28	9		12	2	
28	65		4	3		3			60	33	11		60		
29	64								55		13		24	1	2
30	62								60	100	39		12		1
31	60								62	49	14			2	3
32	58		12	5		2		30	60	35	6	30			
33	52							40	20	30				2	
34	50								30	18	25			1	
35	40								40	39	8		100	1	1
36	37			2			2		33	32					
37	29								29	41	7		240	1	
38†	21								20	24	4		25	1	

†Farm 38 had 1 acre of rough land not in agricultural use

LAND USE

Grassland

Table 14 shows that this is the major type of land use and occurs on all farms in the sample. We can make use of the interval data to estimate the extent to which it is so on all farms in the population. Table 15 shows grassland (permanent and temporary) as a percentage of each farm's acreage. The mean, the best estimate of the standard deviation and the standard error of the mean have been calculated. The population mean lies between 87.5 and 94.7% and the limits of the range of values show that 95% of the farms would have between 68.9 and 100% of their acreage under grass.

Table 15 Farm survey: Grassland as a percentage of all farmland

Farm number	Percentage	$(x-\bar{x})$	$(x-\bar{x})^2$	Farm number	Percentage	$(x-\bar{x})$	$(x-\bar{x})^2$
1	74.5	−16.6	275.56	20	95.4	+4.3	18.49
2	77.7	−13.4	179.56	21	77.1	−14.0	196.00
3	95.3	+4.2	17.64	22	100.0	+8.9	79.21
4	100.0	+8.9	79.21	23	100.0	+8.9	79.21
5	100.0	+8.9	79.21	24	66.7	−24.4	595.36
6	90.0	−1.1	1.21	25	100.0	+8.9	79.21
7	100.0	+8.9	79.21	26	100.0	+8.9	79.21
8	100.0	+8.9	79.21	27	92.3	+1.2	1.44
9	100.0	+8.9	79.21	28	84.6	−6.5	42.25
10	100.0	+8.9	79.21	29	93.7	+2.6	6.76
11	95.7	+4.6	21.16	30	100.0	+8.9	79.21
12	100.0	+8.9	79.21	31	100.0	+8.9	79.21
13	100.0	+8.9	79.21	32	86.2	−4.9	24.01
14	86.7	−4.4	19.36	33	76.9	−14.2	201.64
15	86.8	−4.3	18.49	34	60.0	−31.1	967.21
16	88.9	−2.2	4.84	35	100.0	+8.9	79.21
17	67.5	−23.6	556.96	36	89.2	−1.9	3.61
18	83.3	−7.8	60.84	37	100.0	+8.9	79.21
19	100.0	+8.9	79.21	38	95.2	+4.1	16.81
				Total	3463.7		4575.77
				Mean	91.1		

$$\hat{\sigma} = \sqrt{\frac{\Sigma(x-\bar{x})^2}{n-1}}$$

$$= \sqrt{\frac{4575.77}{37}} = \sqrt{123.7} = 11.1 \qquad \text{SEM} = \frac{\hat{\sigma}}{\sqrt{n}}$$

$$= \frac{11.1}{\sqrt{38}} = 1.8$$

Therefore at the 95% confidence level the population mean lies between 91.1 ± (2 × 1.8), i.e. between 87.5 and 94.7%.

At the same confidence level the range of values for the population lies between 91.1 ± 2$\hat{\sigma}$, i.e. 91.1 ± (2 × 11.1), i.e. between 68.9 and 100%.

Temporary grassland

Though permanent grassland is the predominant type on 30 of the 38 farms and 17 of these have their entire acreage in it, 14 farms have some temporary grassland left down for from 2 to 7 years.

The likely proportion of farms in the total population having some sown grassland can be calculated:

$$p = \frac{14}{38} \times 100 = 36.8\% \therefore q = 63.2\%$$

$$SE = \sqrt{\frac{36.8 \times 63.2}{38}} = \sqrt{61.2} = 7.8$$

At the 95% confidence level the proportion of farms in the area with sown grassland is 36.8 ± (2 × 7.8), i.e. between 21.2 and 52.4%. We can therefore say that only about half the farms are likely to have any sown grassland and the proportion could be as low as one in five.

Arable

Only 15 farms grow crops other than grass.

$$p = \frac{15}{38} \times 100 = 39.5\% \therefore q = 60.5\%$$

$$SE = \sqrt{\frac{39.5 \times 60.5}{38}} = \sqrt{62.9} = 7.9$$

At the 95% confidence level 39.5 ± (2 × 7.9) of farms in the area will grow crops other than grass, i.e. from 23.7 to 55.3%.

It can be seen that the commonest crop is barley. 11 out of 15 grow barley. Only the 2 largest farms grow any wheat and only 3 of the smaller farms grow a few acres of oats. The likelihood of barley being grown can be calculated:

$$p = \frac{11}{15} \times 100 = 73.3\% \therefore q = 26.7\%$$

$$SE = \sqrt{\frac{73.3 \times 26.7}{15}} = \sqrt{130.5} = 11.4$$

This means that at the 95% confidence level the range is 73.3 ± (2 × 11.4), i.e. from 50.5 to 96.1% of the farms growing crops will grow barley. We cannot give more precise estimates at this level of confidence with a number as small as 15 in the sample. If we want to find what proportion of *all* the farms grow barley:

$$p = \frac{11}{38} \times 100 = 28.9\% \therefore q = 71.1\%$$

$$SE = \sqrt{\frac{28.9 \times 71.1}{38}} = \sqrt{54.1} = 7.4$$

Therefore we see that between 14.1 and 43.7% of all farms will be likely to grow barley at the 95% confidence level.

It will be noticed that only one farm in group 2 grows barley, or indeed grows any crop other than grass, and also that the only farms that grow potatoes are among the small farms. It would be highly dangerous to draw conclusions concerning all farms in the area from this data. The one 250 acre farm growing barley represents 10% of the farms in that class. It may be a chance occurrence. We cannot make any precise statements about potatoes or oats on larger farms except to say that none were found in the sample. Of the smaller farms 24% grow potatoes. Applied to the population this could indicate at the 95% confidence level that between 7 and 41% of the small farms grow potatoes. Similar calculations could be worked out for kale and oats but the reader must not lose sight of the purpose of the study. For most purposes it would be more meaningful to say that a few acres of potatoes and oats tend to be grown on the smaller farms, rather than to quote wide confidence limits for these crops. A larger sample from each size group of farms would give more precise results.

Livestock

It is obvious that the mainstay of the farm economy is dairying. Only 2 farms out of the 38 keep no dairy cattle: these rear cattle for beef. For dairy cattle:

$$p = \frac{36}{38} \times 100 = 94.7\% \therefore q = 5.3\%$$

$$SE = \sqrt{\frac{94.7 \times 5.3}{38}} = \sqrt{13.2} = 3.6$$

At the 95% confidence level the range of values is 87 to 100%, showing that probably at least 87% of all farms keep dairy cattle.

Similar calculations could be made for the proportion of farms with other types of livestock. It is clear that sheep tend to be absent from the smaller farms but poultry are reared on farms of various sizes, though not on all farms.

Having established the extent to which the dominant farm practice in the area is grassland dairy farming supplemented by other enterprises which vary to some extent with the size of farm, it is worth considering the ratio of cattle to farm acreage. Table 16 shows the number of cattle per 100 acres (calculated to the nearest whole number) on each farm. For example farm No. 1 has 403 cattle (dairy and others) and 1100 acres of land, i.e. $\frac{100 \times 403}{1100} = 37$ cattle per 100 acres. The mean, best estimate of the standard deviation and standard error of the mean are shown on Table 16. Figure 12 shows graphically the distribution of values round the sample mean of 63. The population mean lies between

53 and 73 at the 95% confidence level and the limits of the range of values show that 95% of the farms in the area would keep between 3 and 123 cattle per 100 acres of farm land.

Table 16 Farmland survey: Cattle per 100 acres (to nearest whole number)

Farm number	Number of cattle	$(x-\bar{x})$	$(x-\bar{x})^2$	Farm number	Number of cattle	$(x-\bar{x})$	$(x-\bar{x})^2$
1	37	−26	676	20	56	−7	49
2	20	−43	1849	21	57	−6	36
3	43	−20	400	22	81	+18	324
4	45	−18	324	23	40	−23	529
5	69	+6	36	24	57	−6	36
6	17	−46	2116	25	43	−20	400
7	38	−25	625	26	95	+32	1024
8	64	+1	1	27	57	−6	36
9	59	−4	16	28	68	+5	25
10	32	−31	961	29	97	+34	1156
11	44	−19	361	30	63	0	0
12	54	−9	81	31	82	+19	361
13	31	−32	1024	32	62	−1	1
14	69	+6	36	33	35	−28	784
15	35	−28	784	34	128	+65	4225
16	100	+37	1369	35	100	+37	1369
17	25	−38	1444	36	111	+48	2304
18	68	+5	25	37	107	+44	1936
19	142	+79	6241	38	57	−6	36
				Total	2387		33 000
				Mean	63		

$$\hat{\sigma} = \sqrt{\frac{\Sigma(x-\bar{x})^2}{n-1}}$$

$$= \sqrt{\frac{33\,000}{37}} = \sqrt{891.1} = 29.9$$

$$\text{SEM} = \frac{\hat{\sigma}}{\sqrt{n}}$$

$$= \frac{29.9}{\sqrt{38}} = 4.8$$

Therefore at the 95% confidence level the average number of cattle per 100 acres on all farms in the area is 63 ± (2 × 4.8), i.e. between 53 and 73 to the nearest whole number.

At the same confidence level the range of values for the population lies between 63 ± 2$\hat{\sigma}$, i.e. 63 ± (2 × 4.8), i.e. between 3 and 123 to the nearest whole number.

With the credentials of the sample established, it may now be of interest to look at the extremes of the distribution. The two farms which keep relatively few cattle engage more than others in other enterprises, since No. 2 keeps a much larger flock of sheep than any other farm and No. 6 keeps a much larger

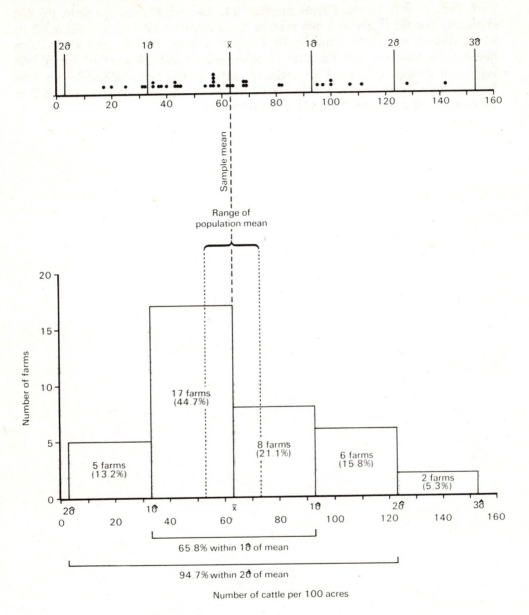

Figure 12 Distribution of sample farms according to cattle density

flock of poultry. Those with relatively large numbers of cattle per acre tend to grow few, if any, crops. Farms number 34, 35 and 37, for example, do not grow any fodder crops or sown grasses, farm number 19 has only 33 acres of sown grasses and farm number 36 has just 4 acres of oats and kale. Farm number 16, however, has 60 acres of sown grasses and 14 acres of barley and kale even though it has a high ratio of cattle to the acre. Sampling can thus lead to the identification, within stated limits, of the general farm practice in an area and of deviation from the general pattern.

CHAPTER ELEVEN

Stratified sampling in settlement study

An evaluation of the facilities available in settlements in a rural area is made in this example to demonstrate stratified sampling techniques.

Method

In first-hand study in the field, shopping facilities might be selected, or a more comprehensive evaluation can be attempted. A useful method of quantifying facilities so that one settlement can be compared with another is to allot points for each facility present.

The following scale is a suggestion of how this may be done. Other facilities may be added if they are relevant and the system can be modified to accommodate local characteristics.

Three points: Secondary school, bus service every two hours, village hall.

Two points: Daily bus service, doctor's surgery, district nurse, sewerage system, each industry employing more than five people, playing field.

One point: Church, chapel, primary school, clinic, library service, post office, each shop, public house, garage, public water supply, electricity, gas, bus service less than daily, each village organisation, e.g. Scouts, Women's Institute.

The problem considered here is to ascertain the point scores in an area containing approximately 100 settlements. Time would limit the number which could be visited to collect the information. A carefully organised sample survey would enable an estimate to be made within stated confidence limits.

Selecting the sample

It can be assumed that the biggest factor in determining the number of facilities is size of settlement. A large population will support more shops and organizations than a small one. In an area where settlements vary much in size overall random or systematic methods could result in under representation of settlements of a certain size. Stratification of the settlements (the 'population') based on number of people would ensure that settlements of all size groups are adequately represented. Also estimates for settlements of different sizes would be possible.

The first step is to arrange the settlements in size groups. This can be done by obtaining the population figures from the County Census Report and plotting them on a scatter diagram. Figure 13a shows the 98 settlements in part of Lincolnshire. They vary in size from 11 000 to 22 people. Careful consideration should be given to the best way to group them into strata. Some knowledge of the area is presumed; the six largest settlements need separate treat-

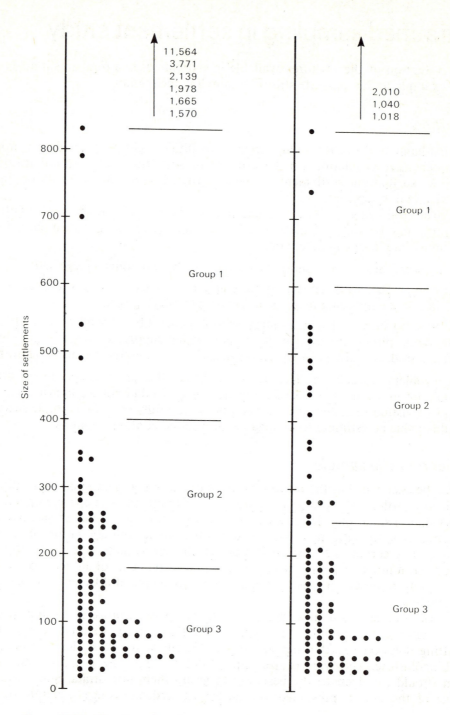

Figure 13 Stratification of settlements

ment as they are service centres for a wider area than that within their bounds. They will be excluded from the treatment here.

The remainder range in size from 22 to 830 inhabitants. Most of them fall into the range of 22 to 400. A decision must be made as to where to divide this large group and into how many strata. Three strata in all are used in this analysis:

Group	Number of inhabitants	Number of settlements
1	401–830	5
2	180–400	27
3	less than 180	60
		Total 92

A sample is needed for each stratum. The next problem is to decide how many from each. At least 30 overall would be advisable if time and manpower permit. Proportional representation, that is using a uniform sample fraction, would give:

Group	Population	Sampling fraction	Sample size*
1	5	30/92	2
2	27	30/92	9
3	60	30/92	20
Total	**92**		**31**

*To nearest whole number

This has grave disadvantages. A sample of 2 is too small. Consider a chance occurrence of one of the settlements having either fewer, or many more, facilities than the others in the group for some reason. This village, if selected in the sample, would represent 50% of it. When the sample results were raised to apply to the population this chance occurrence would carry undue weight. If it were one of a large sample it would carry a weighting closer to the correct value. In fact the population of 5 is too small for sampling and all units should be included in the study.

There remain then two groups of 27 and 60 units respectively from which to select 25 in all. Proportional representation might be considered for these two groups:

$$\text{Group 2: } \frac{25}{87} \times 27 = 8^* \qquad \text{Group 3: } \frac{25}{87} \times 60 = 17^*$$

*To the nearest whole number

Again 8 only from group 2 is a small number and if the range of scores is wide (i.e. the group has a large standard deviation) the standard error of the mean will be too large to allow sufficiently precise statements to be made about the population. Previous experience in rural settlement studies may help in making a decision. It is likely that facilities will vary more in group 2 than in group 3, so a larger proportion would be needed to give comparable standard errors. A better use of resources might be to take 10 from group 2 and 15 from group 3. This is termed using a variable sampling fraction (f), which will mean a variable raising factor.

$$\text{Group 2:} \quad f = \frac{10}{27} = 0.37 \qquad \text{Group 3:} \quad f = \frac{15}{60} = 0.25$$

It can be seen that careful thought is needed in working out the allocation of time and manpower to the best advantage. It might in fact be advisable to take a small sample to get an idea of the number likely to be necessary to achieve estimates suitable for the study, if the area is not well known.

Another instance of the problem is shown on Figure 13b. This shows eighty settlements of comparable size to those considered above, in another part of Eastern England. The settlements range from 30 to 2010 population. If it were desired to use stratified random sampling methods for this area a decision would need to be made as to how many strata and where to draw the lines. Again the three largest settlements stand alone. The three of 600–830 would form a group which would need total coverage. For the rest a break could be made at 250 giving 3 strata in all.

Group	Number of inhabitants	Number of settlements
1	600–830	3
2	250–530	16
3	less than 250	58
		Total 77

A knowledge of settlements in the area might suggest that the smaller villages differ little in their amenities and a sample of 10 from group 3 would be likely to give results comparable to those obtained from a sample of 10 from group 2. Thus using a variable sampling fraction would give:

Group	Population	Sampling fraction	Sample size
1	3	3/3	3*
2	16	10/16	10
3	58	10/58	10
Total	77		23

*Total coverage

The statistics of the first example considered above (*Figure 13a*), using a variable sampling fraction, will be used to illustrate the procedures which follow. The settlements in each group should be listed in alphabetical order to avoid bias and then numbered. Group 2 would be numbered 1 to 27 and group 3, 1 to 60. The requisite number for the sample is selected from each of these groups separately by random number method, 10 from group 2 and 15 from group 3. These are the settlements to be visited, their facilities recorded and points allotted.

Tabulation and processing

Table 17 shows the tabulated scores. The mean and the standard deviation have then been calculated for each group. For group 1 these are the population parameters as the whole of the population has been studied. For groups 2 and 3 the mean is a sample parameter and the standard deviation is the best estimate of the standard deviation of the population. It is necessary to calculate the standard error of each of the means.

Here a modification to the formula which has already been learnt for standard error must be introduced. When the population is small and the sample is large in proportion, $\hat{\sigma}$ is very near to σ and the standard error of the mean is almost always less than the formula $\frac{\hat{\sigma}}{\sqrt{n}}$ would suggest. This formula can be modified in stratified sampling by the factor $\sqrt{(1-f)}$, where f is the sampling fraction. This can best be appreciated by considering that if all the population were included in the survey f would be 1 (total coverage); $\sqrt{(1-f)}$ would be equal to 0 and the standard error of the mean would also be zero, since:

$$\frac{\hat{\sigma}}{\sqrt{n}} \times \sqrt{(1-1)} = 0.$$

Table 17 Rural settlement survey: Summary of data

GROUP 1

Population: 5 villages

Village number	Points	$(x-\bar{x})$	$(x-\bar{x})^2$
1	43	9.8	96.04
2	35	1.8	3.24
3	26	−7.2	51.84
4	30	−3.2	10.24
5	32	1.2	1.44
Total	166		162.80
Mean	33.2		

$$\sigma = \sqrt{\frac{\Sigma(x-\bar{x})^2}{n}}$$

$$= \sqrt{\frac{162.8}{5}} = \sqrt{32.6} = 5.7$$

GROUP 2

Population: 27 villages; 10 in sample

Village number	Points	$(x-\bar{x})$	$(x-\bar{x})^2$
17	14	0.4	0.16
5	9	−4.6	21.16
23	16	2.4	5.76
2	20	6.4	40.96
1	14	0.4	0.16
21	14	0.4	0.16
4	10	−3.6	12.96
14	20	6.4	40.96
26	8	−5.6	31.36
8	11	−2.6	6.76
Total	136		160.40
Mean	13.6		

$$\hat{\sigma} = \sqrt{\frac{\Sigma(x-\bar{x})^2}{n-1}}$$

$$= \sqrt{\frac{160.4}{9}} = \sqrt{17.8} = 4.2$$

$$\text{SEM} = \frac{\hat{\sigma}}{\sqrt{n}} \times \sqrt{(1-f)}$$

$$f = 0.37$$

$$\therefore \text{SEM} = \frac{4.2}{\sqrt{10}} \times \sqrt{0.63} = 1.0$$

The population mean is therefore in the range 13.6 ± ($t \times$ 1.0) at the 95% confidence level, where t = 2.3 with 9 degrees of freedom. Therefore the population mean lies in the range 13.6 ± (2.3 × 1.0), i.e. from 11.3 to 15.9.

At the same confidence level the range of points for all the villages in the group lies between 13.6 ± ($t \times \hat{\sigma}$), i.e. 13.6 ± (2.3 × 4.2), i.e. between 4 and 23 points to the nearest whole number.

GROUP 3

Population: 60 villages; 15 in sample

Village number	Points	$(x-\bar{x})^2$	$(x-\bar{x})^2$
39	15	5.3	28.09
27	11	1.3	1.69
40	8	−1.7	2.89
33	5	−4.7	22.09
36	10	0.3	0.09
59	15	5.3	28.09
20	5	−4.7	22.09
55	9	−0.7	0.49
41	10	0.3	0.09
13	8	−1.7	2.89
7	13	3.3	10.89
30	14	4.3	18.49
9	6	−3.7	13.69
14	11	1.3	1.69
5	6	−3.7	13.69
Total	146		166.95
Mean	9.7		

$$\hat{\sigma} = \sqrt{\frac{167.0}{14}}$$

$$= \sqrt{11.9} = 3.5$$

$$f = 0.25$$

$$\therefore \text{SEM} = \frac{3.5}{\sqrt{15}} \times \sqrt{0.75} = 0.8$$

The population mean is therefore in the range 9.7 ± ($t \times$ 0.8) at the 95% confidence level, where t = 2.1 with 14 degress of freedom. Therefore the population mean lies in the range 9.7 ± (2.1 × 0.8), i.e. from 8.0 to 11.4.

At the same confidence level the range of points for all the villages in the group lies between 9.7 ± (t × ô), i.e. 9.7 ± (2.1 × 3.5), i.e. between 2 and 17 points to the nearest whole number.

The parameters of the strata can be tabulated as follows:

Group	Number in sample	Sample mean	Standard error of the mean	Number in population	Range of population mean	Standard deviation	Range of point scores
1	5			5	33.2	5.7	26–43*
2	10	13.6	1.0	27	11.3–15.9	4.2	4–23†
3	15	9.7	0.8	60	8.0–11.4	3.5	2–17†

*Actual values   †Estimated values

To obtain a value of one for the standard error in group 3 a sample of 12 would have been sufficient. For group 1 the tabulated results are the population parameters since the 'sample' consists of the whole population and the range of points shown is the actual range taken from the survey results.

Several points of interest emerge. The largest standard deviation shows that the largest villages are more varied in the facilities they have to offer, we can say this with complete confidence. Some of them may well serve smaller settlements in their areas for higher order facilities such as doctor's surgery, district nurse, small industry for employment and primary schooling. A further study could be made of their catchment areas.

Group 3 is the most homogeneous, showing by the smallest standard deviation that the smallest villages vary relatively little in their facilities. Many lines of further enquiry are opened. The most important point to be emphasised here is the method of selecting and processing the strata data.

It is now possible to estimate the overall provision of facilities in all settlements of less than 1000 people. This must be done for each stratum separately and then the stratum totals can be summed. This procedure is necessary because a variable sampling fraction has been used.

The procedure will be clear if the information gained from the samples is entered on to a table such as Table 18 and the population parameters recorded as they are calculated. Those for group 1 have already been obtained during the initial survey.

For the other groups the total sample number of facilities must be multiplied by the raising factor (rf) for the group to find the population total. The raising factor is the reciprocal of the sampling fraction (f).

$$\text{Group 2: } f = \frac{10}{27} \quad \therefore rf = \frac{27}{10} = 2.7$$

$$\therefore \text{Population total} = 136 \times 2.7 = 367$$

Table 18 Rural settlement survey: Summary of results

Group	Number of inhabitants	SAMPLE					
		Number of points in sample	Sampling fraction	Raising factor	Total points in sample	Sample mean	Standard error of mean
1	401–800	5	1.0	1.0	166	33.2	0
2	180–400	10	0.37	2.7	136	13.6	1.0
3	less than 180	15	0.25	4.0	146	9.7	0.8
							0.6

$$\text{Group 3: } f = \frac{15}{60} \quad \therefore rf = \frac{60}{15} = 4.0$$

$$\therefore \text{Population total} = 146 \times 40 = 584$$

The overall mean is the total number of points for settlements in the three groups divided by the overall number of units in the groups:

$$166 + 367 + 584 = 1117$$

$$\therefore \text{Overall mean} = \frac{1117}{92} = 12.1$$

The population totals can be used as an index for the area and to compare it with other rural areas, provided we know the level of confidence with which we can accept them. Their standard errors must be calculated and in doing this the different raising factors must again be taken into account. All the values concerned in the calculations have already been obtained. The *standard error of the estimated stratum total* is obtained by substituting in the formula

$$rf \sqrt{[\hat{\sigma}^2 \times n(1-f)]}$$

where rf = raising factor for a certain stratum
$\hat{\sigma}$ = best estimate of the standard deviation
n = number of units in the sample
f = sampling fraction for the group

This can be written as $\sqrt{[\hat{\sigma}^2 \times n(1-f) \times (rf)^2]}$

	POPULATION				
Number of units in stratum	Total points in stratum	Stratum mean	Standard deviation	Standard error of totals	Group
5	166	33.2	5.7	0	1
27	367 est.	11.3 to 15.9	4.2 est.	28.4	2
60	584 est.	8.0 to 11.4	3.5 est.	47.0	3
92	1117 est.	12.1	5.8	54.9	Overall

Group 1:

$$rf\sqrt{[\hat{\sigma}^2 \times n(1-f)]} = 1\sqrt{[(5.7)^2 \times 5(1-1)]}$$
$$= 0$$

Group 2:

$$rf\sqrt{[\hat{\sigma}^2 \times n(1-f)]} = 2.7\sqrt{[(4.2)^2 \times 10(1.0-0.37)]}$$
$$= 2.7\sqrt{111.1} = 28.4$$

Therefore at the 95% confidence level the total number of points for villages of 180 to 400 inhabitants lies between 367 ± (2.3 × 28.4), i.e. between 302 and 432 to the nearest whole number.

Group 3:

$$rf\sqrt{[\hat{\sigma}^2 \times n(1-f)]} = 4.0\sqrt{[(3.5)^2 \times 15(1.0-0.25)]}$$
$$= 4.0\sqrt{137.8} = 47.0$$

Therefore at the 95% confidence level the total number of points for villages of less than 180 inhabitants lies between 584 ± (2.1 × 47.0), i.e. between 485 and 683 to the nearest whole number.

The *standard error of the estimated overall population total* is the square root of the sum of the squares of the individual standard errors of the estimated stratum totals calculated above:

$$\sqrt{\Sigma[\hat{\sigma}^2 \times n(1-f) \times (rf)^2]} = \sqrt{[(0)^2 + (28.4)^2 + (47.0)^2]}$$
$$= \sqrt{3015.6} = 54.9$$

The *standard error of the estimated overall sample mean* is obtained by dividing the standard error of the estimated overall population total, calculated above, by the overall number of units in the groups:

$$\frac{\sqrt{\Sigma[\hat{\sigma}^2 \times n(1-f) \times (rf)^2]}}{N} = \frac{54.9}{92} = 0.6$$

where N = overall number of units in the groups

From the formula SEM = $\frac{\hat{\sigma}}{\sqrt{n}}$, the *overall standard deviation* is:

$$\text{SEM} \times \sqrt{N} = 0.6 \times \sqrt{92} = 5.8$$

Thus from the overall standard deviation (5.8) and the overall mean (12.1) it can be said that 95% of the settlements in the area have scores of 12.1 ± (2 × 5.8), i.e. between 0.5 and 23.7.

A sample study such as this could be the starting point for detailed investigation of the relationship between size and facilities in rural settlements. It could also provide indices with which to compare one area with another in a different part of the country. A close assessment of the characteristics of a large number of settlements can be obtained from a survey of 30 or so of them provided the sampling is efficiently organised and the findings properly processed.

III — COMPARISON OF SAMPLE VALUES

CHAPTER TWELVE

Significance of sample differences for interval data

Geographical work frequently involves the comparison of areas and the grouping together of similar areas to delimit formal regions. It is highly dangerous to compare areas for which sample data only is available without first ascertaining whether the differences between the samples are statistically significant, such that they justify a statement that the population means differ. For interval data, if the differences are statistically significant and not the result of chance selection the differences will still occur between other samples taken from the same populations and will be greater than the differences within the populations themselves. In the case of frequency data detailed in the following chapter, if the differences are real, the observed frequencies will differ significantly from the expected frequencies.

If two samples are to be compared it is necessary to calculate the standard error of the differences between their mean values. The parameters of both samples, that is, their means and their standard deviations, are involved in the calculation. Data from examples worked in Part II of this book will be used to illustrate the method.

In the example in Chapter 8 for the sample from the village of Welton, the mean number of mentions of Lincoln as a service centre for the village was 14.8 per household. For the sample from Ingham it was 13.9. The problem is to ascertain at a stated level of confidence whether Welton households do in fact use Lincoln services to a greater extent than Ingham households, or whether the small difference shown in the sample means should be discounted because it could have occurred as a result of chance in the selection of the particular units of the sample.

The formula for the standard error of the difference between the two sample means, $\bar{a}$ and $\bar{b}$, is

$$\sqrt{(SEM_a^2 + SEM_b^2)}$$

where $SEM = \dfrac{\hat{\sigma}}{\sqrt{n}}$

$$\text{i.e. } SE(\bar{a}-\bar{b}) = \sqrt{\left(\frac{\hat{\sigma}_a^2}{n_a} + \frac{\hat{\sigma}_b^2}{n_b}\right)}$$

In this example the letters a and b denote the samples from Welton and Ingham.

$$SE(\bar{a}-\bar{b}) = \sqrt{\left(\frac{22.2}{30} + \frac{9.9}{20}\right)}$$

$$= \sqrt{1.2} \quad = 1.1$$

	n	Mean	$\hat{\sigma}^2$
a Welton	30	14.8	22.2
b Ingham	20	13.9	9.9

If the number of units in the two samples combined is more than 30 the 'two standard error' method can be used to state the meaning of this error. At the 95% confidence level, if the actual difference of the sample means is more than twice the standard error of the difference, then it is unlikely that the difference between the two sample means has occurred by chance. Therefore this difference is significant and a difference would occur between the population means. If the difference is greater than three times the standard error of the difference one can say that it is a real difference at the 99% confidence level.

In the above example:

$$\bar{a}-\bar{b} = 14.8 - 13.9 = 0.9$$

$$2 \times SE(\bar{a}-\bar{b}) = 2.2$$

$(\bar{a} - \bar{b})$ is less than $2 \times SE(\bar{a} - \bar{b})$, so the difference between the two sample means is not significant and a real difference has not been established. The sample values should not be used for comparing conditions in the two villages. We cannot say definitely that there is no difference, but we must say that none has been proved by the sample survey. If it were necessary to pursue the matter further a larger sample would have to be taken to attempt to prove that real differences exist. In this case however the $(\bar{a} - \bar{b})$ value is so much below the value of twice its standard error that a real difference is very unlikely to exist.

In Chapter 9 in the example of the Derbyshire villages, it was suggested that the reason for taking the samples was to map the relative volumes of movements between different villages and their main service centres. The samples should be used for this purpose only if the differences between them are significant. For example:

$$\bar{a}-\bar{b} = 6.7 - 4.9 = 1.8$$

SIGNIFICANCE OF SAMPLE DIFFERENCES FOR INTERVAL DATA

$$SE(\bar{a}-\bar{b}) = \sqrt{\left(\frac{7.4}{30} + \frac{6.8}{30}\right)}$$

$$= \sqrt{0.47} = 0.7$$

$$2 \times 0.7 = 1.2$$

	n	Mean	$\hat{\sigma}^2$
a Elton	30	6.7	7.4
b Beeley	30	4.9	6.8
c Bonsall	30	12.2	17.9

Since the difference in the two sample means (1.8) is greater than twice the standard error of their difference (1.2) the difference between the sample means is significant at the 95% confidence level.

$$\bar{c}-\bar{b} = 12.2 - 4.9 = 7.3$$

$$SE(\bar{c}-\bar{b}) = \sqrt{\left(\frac{17.9}{30} \times \frac{6.8}{30}\right)}$$

$$= \sqrt{0.8} = 0.9$$

$$2 \times 0.9 = 1.8$$

$$3 \times 0.9 = 2.7$$

Since 7.3 is greater than twice $SE(\bar{c}-\bar{b})$, the difference between $\bar{c}$ and $\bar{b}$ is significant at the 95% confidence level. But 7.3 is greater than 3×0.9 and therefore the difference between $\bar{c}$ and $\bar{b}$ is also significant at the 99% confidence level.

$$\bar{c}-\bar{a} = 17.9 - 12.2 = 5.5$$

$$SE(\bar{c}-\bar{a}) = \sqrt{\left(\frac{17.9}{30} + \frac{7.4}{30}\right)}$$

$$= \sqrt{0.8} = 0.9$$

$$2 \times 0.9 = 1.8$$

Since 5.5 is greater than 2×0.9 the difference between $\bar{c}$ and $\bar{a}$ is significant at the 95% confidence level.

It is in fact so for all five villages and the larger movements shown by sample survey are mapped on Figure 14.

When the samples are small, less than about 30 units, Student's t test should be used instead of the 'two standard error' test for significance.

$$\text{The index } t = \frac{\text{difference between two means}}{\text{standard error of the difference}}$$

$$= \frac{\bar{a}-\bar{b}}{SE(\bar{a}-\bar{b})}$$

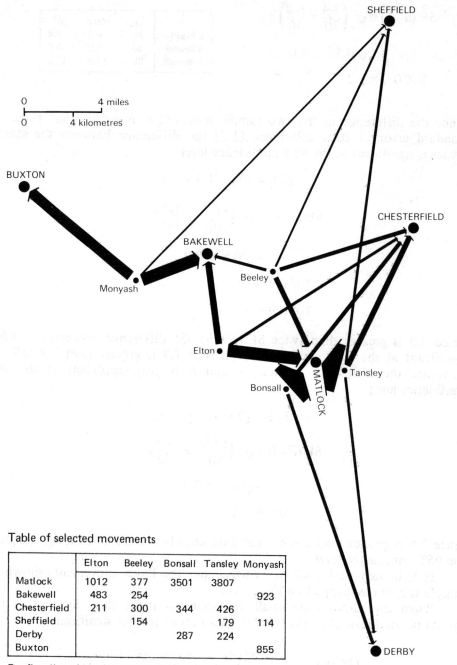

Figure 14 Flow diagram of selected movements for goods and services

SIGNIFICANCE OF SAMPLE DIFFERENCES FOR INTERVAL DATA

In the example in Chapter 8:

$$\bar{a} - \bar{b} = 4.4$$

$$\text{SE}(\bar{a} - \bar{b}) = \sqrt{\left(\frac{26.9}{7} + \frac{11.9}{7}\right)}$$

$$= \sqrt{5.5} = 2.3$$

	n	Mean	$\hat{\sigma}^2$
a Hackthorn	7	15.7	26.9
b Fillingham	7	11.3	11.9

$$t = \frac{4.4}{2.3} = 1.9$$

The number of degrees of freedom is $(n_a - 1) + (n_b - 1) = 12$. The significance level for $t = 1.9$ with 12 degrees of freedom must be looked up on the t tables (*Appendix 4*) or t graph. It lies between 10 and 5%. This means that there is more than 5% probability that the difference has occurred by chance and it is not proved to be significant at the 95% confidence level.

In the example of stratified sampling in Chapter 11:

$$\bar{a} - \bar{b} = 3.9$$

$$\text{SE}(\bar{a} - \bar{b}) = \sqrt{\left(\frac{17.8}{10} + \frac{11.9}{15}\right)}$$

$$= \sqrt{2.6} = 1.6$$

	n	Mean	$\hat{\sigma}^2$
a Group 2 villages	10	13.6	17.8
b Group 3 villages	15	9.7	11.9

$$t = \frac{3.9}{1.6} = 2.44$$

The number of degrees of freedom is $(n_a - 1) + (n_b - 1) = 23$. The significance level for t with 23 degrees of freedom lies between 5 and 2% meaning that the difference in the sample means is very unlikely to be a chance occurrence and is significant above the 95% confidence level. We can say therefore at this level of confidence that all settlements of below 180 people have on average a lower count of facilities than those of between 180 and 400 people and that the difference is in the ratio of 14 to 10, the standard errors of these means being only *a*) 1.0 and *b*) 0.8 (*see summary, page 75*).

The statistical significance of the difference between sets of sample data must be established before the data can be used for further work. If a difference is proved to be significant at least at the 95% confidence level, this means that there is less than 5% probability of it being a chance occurrence. One may then go on to enquire into the reason for the difference, or the possible consequences of the difference. In applied geography a sample survey of existing conditions, provided it is efficiently executed and processed, can supply facts from which local or regional deficiencies may be spotted and future policy developed.

CHAPTER THIRTEEN

Significance of sample differences for frequency data

With sample means derived from frequency data as in the land use examples in Part I the significance of the differences between sample values must be established by use of the Chi-squared test. The reader may be familiar with this test in a different context. In the first book of this series[1] the use of Chi-squared for comparison and, by implication, correlation was discussed.

The Chi-squared number tests whether observed frequencies differ significantly from the frequencies which would be expected according to an assumed hypothesis. In the case of the land use data from the stratified sample in Chapter 4 this would be the frequencies expected if the distribution of each type of land use were proportional to the areas of the two types of land, upland and lowland. For example if a quarter of the area were lowland one would expect a quarter of the total amount of grassland to be on the lowland if the hypothesis held true.

It is essential first to set up the null hypothesis that there is no significant difference in the proportions of land use which occur on the lowland and upland parts of the area. The formula for Chi-squared (χ^2) is:

$$\Sigma \frac{(O-E)^2}{E}$$

where O = the observed values
      E = the values to be expected if the null hypothesis holds true

We are dealing with the two sets of frequency data together — grassland and arable. The observed frequencies should be tabulated as follows:

	A	G	Total
Upland	20	117	137
Lowland	96	87	183
Total	116	204	320

The upland forms $\frac{137}{320}$ of the total area, therefore the occurrence of grassland and arable would be expected to be this fraction of the total occurrence in the whole area.

Upland grassland: $E = \frac{137}{320} \times 204 = 87.3$

---

[1] Garlick, E. J. et al, *Correlation Techniques in Geography*, Philip, 1972

Upland arable: $$E = \frac{137}{320} \times 116 = 49.7$$

The lowland forms $\frac{183}{320}$ of the total area:

Lowland grassland: $$E = \frac{183}{320} \times 204 = 116.7$$

Lowland arable: $$E = \frac{183}{320} \times 116 = 66.4$$

The number of degrees of freedom for the allocation of points must be calculated. Points may be allocated either in upland or lowland; the number of possible categories is 2, therefore $n - 1 = 1$. Points may be either grassland or arable; the number of possible categories is 2, $\therefore n - 1 = 1$. The total number of degrees of freedom is the product of $(n - 1)(n - 1) = 1$. With only 1 degree of freedom it is necessary to apply Yates's correction by adjusting the observed values by 0.5 in the direction of the expected values. Yates's correction gives:

		A	G	Totals
Upland	O	20.5	116.5	137.0
	E	49.7	87.3	137.0
Lowland	O	95.5	87.5	183.0
	E	66.4	116.7	183.0
Totals	O	116.0	204.0	320.0
	E	116.0	204.0	320.0

The value of Chi-squared is calculated in the following table:

	Upland		Lowland		Total
	A	G	A	G	
O	20.5	116.5	95.5	87.5	320.0
E*	49.7	87.3	66.4	116.7	320.1
O−E	−29.2	29.2	29.1	−29.2	
(O−E)²	852.6	852.6	846.8	852.6	
(O−E)²/E	17.2	9.8	12.8	7.3	
χ^2 =	17.2 +	9.8	+ 12.8 +	7.3	= 47.1

*Correct to one decimal place

The χ^2 value of 47.1 with 1 degree of freedom must be looked up on the Chi-squared table (*Appendix 3*). A probability of less than 0.001 is shown. This means that the hypothesis can be accepted with only 0.001 probability. In other words it must be rejected and statistical support for real difference is given. Thus the observed difference in proportions of land use on upland and lowland is significant at the 99.9% confidence level.

The sample shopping survey discussed in the example in Chapter 7 could be a starting point for further work on the central place function of Lincoln, taking into account the number of potential shoppers in the hinterland, accessibility and other factors. The question then arises as to whether the differences found in the sample between numbers from the three distance zones are significant, or whether they are due to chance in the selection of shoppers to question. The null hypothesis might state that there is no difference in the number of shoppers from the three zones. If this were so we could expect the same proportion of the total from each zone, that is a third of 116 from each. The number of degrees of freedom is $(3-1) = 2$, therefore Yates's correction is not necessary.

	8km	8–16km	Over 16km	Total
O	63	38	15	116
E^*	39	39	39	117
$O-E$	24	−1	−24	
$(O-E)^2$	576	1	576	
$\frac{(O-E)^2}{E}$	14.8	0	14.8	

$$x^2 = 14.8 + 0 + 14.8 = 29.6$$

*To nearest whole number

With 2 degrees of freedom this number is greater than that necessary for a probability of 0.001. The probability of there being no difference is therefore less than 1 in 1000. The null hypothesis can be rejected at the 99.9% confidence level.

This having been established the sample results can be used for further work in seeking for an explanation of them. In this case distance, accessibility, population distribution and the location of other centres can be considered as factors contributing to the differences. Further work might, for example, be concerned with correlation between population and numbers using Lincoln for shopping, or the sample results could be considered in relation to theoretical breakpoints calculated between Lincoln and other accessible centres.

The point to be made here is that before sample data can be used for such purposes the basic premise must be proved valid.

Conclusion

Sampling is not the answer to all coverage problems. However it can enable conclusions concerning a large body of data to be reached at a stated confidence level, from a study of only part of the data. The degree of accuracy of such conclusions will depend upon correct selection of that part and accurate processing and interpretation of the results. Sampling enables reliable estimates to be made when the task of studying the whole body of data would be impossible either because the data is infinite, or because time and resources do not permit. In dealing with published statistics where the whole body of data is available similar techniques to those demonstrated in Part II can be used, but time and effort can be wasted if sample data prove not to be significant at the end of the processing. Time might have been better spent on the whole body of data.

Perhaps the most important consideration in sampling is the integrity of the worker in carrying out procedures and drawing conclusions. Further work derived from unproven samples is useless. If geographers are to take quantification seriously the limitations of the data must be understood. Any of the examples worked here, with their limits of significance is sufficient to demonstrate the dangers of assuming precision in a survey when in fact no such assumption is justified. Sampling procedures and what they reveal are essential to anyone seeking familiarity with modern techniques in geography.

SELECTED BIBLIOGRAPHY

Alder, H. L., and E. B. Roessler, *Introduction to Probability and Statistics*, Freeman, 1972
Berry, B. J. L., and D. Marble (eds.), *Spatial Analysis*, Prentice-Hall, 1968
Cole, J. P., and C. A. M. King, *Quantitative Geography*, Wiley, 1968
Conway, F., *Descriptive Statistics*, Leicester University Press, 1967
Franzblau, A. N., *A Primer of Statistics for non-Statisticians*, Harcourt Brace Jovanovich, 1974
Gregory, S., *Statistical Methods and the Geographer*, Longman, 1973
Haggett, P., *Locational Analysis in Human Geography*, Arnold, 1965
Harvey, D., *Explanation in Geography*, Arnold, 1969
King, L. J., *Statistical Analysis in Geography*, Prentice-Hall, 1969
Moser, C. A., *Survey Methods in Social Investigation*, Heinemann Educational, 1971
Siegel, S., *Nonparametric Statistics for Behavioural Science*, McGraw-Hill, 1956
Stuart, A., *Basic Ideas of Scientific Sampling*, Griffin, 1962
Yates, F., *Sampling Methods for Censuses and Surveys*, Griffin, 1960

APPENDIX I

Summary of formulae

Frequency data

1) *Standard error of sample frequencies:* $SE = \sqrt{\dfrac{p \times q}{n}}$

   where p = the percentage of occurrences in a specified category
   q = the percentage not in that category ($\therefore q = 100 - p$)
   n = the number of points in the sample

2) *Comparison of samples:* Chi-squared $(\chi^2) = \Sigma \dfrac{(O - E)^2}{E}$

   where O = the observed values
   E = the expected values according to a stated null hypothesis

   The Chi-squared (χ^2) test is used to establish the significance of differences in sample values.

3) *Size of sample (n) required to give a desired standard error (d):* $n = \dfrac{p \times q}{d^2}$

Interval data

1) *Sample mean:* $\bar{x} = \dfrac{\Sigma x}{n}$

   where x = the individual values in the sample
   n = the number in the sample

2) *Best estimate of the standard deviation:* $\hat{\sigma} = \sqrt{\dfrac{(x - \bar{x})^2}{n - 1}}$

3) *Standard error of the mean:* $SEM = \dfrac{\hat{\sigma}}{\sqrt{n}}$

4) *Size of sample (n) required to give a desired standard error (d):* $n = \left(\dfrac{\hat{\sigma}}{d}\right)^2$

5) *Stratified sampling:*

   Standard error of the stratum sample mean $= \sqrt{\left[\dfrac{\hat{\sigma}^2}{n} \times (1 - f)\right]}$

   Standard error of the stratum population total
   $= rf\sqrt{[\hat{\sigma}^2 \times n(1 - f)]}$ or $\sqrt{[\hat{\sigma}^2 \times n(1 - f) \times (rf)^2]}$

   Standard error of the overall population total $= \sqrt{\Sigma[\hat{\sigma}^2 \times n(1 - f) \times (rf)^2]}$

   Standard error of the overall sample mean $= \dfrac{\sqrt{\Sigma[\hat{\sigma}^2 \times n(1 - f) \times (rf)^2]}}{N}$

   where f = the sampling fraction
   rf = the raising factor
   n = the number of units in each stratum
   N = the overall number of units

6) *Comparison of samples:*

The standard error of the difference between the two sample means $\bar{a}$ and $\bar{b}$ [SE $(\bar{a} - \bar{b})$]

$$= \sqrt{\left(\frac{\hat{\sigma}_a^2}{n_a} \times \frac{\hat{\sigma}_b^2}{n_b}\right)}$$

where a and b = the two samples being compared

Student's $t = \dfrac{\text{difference between sample means}}{\text{standard error of the difference}}$

$= \dfrac{\bar{a} - \bar{b}}{\text{SE}(\bar{a} - \bar{b})}$

Random sampling numbers

APPENDIX 2

20 17	42 28	23 17	59 66	38 61	02 10	86 10	51 55	92 52	44 25
74 49	04 49	03 04	10 33	53 70	11 54	48 63	94 60	94 49	57 38
94 70	49 31	38 67	23 42	29 65	40 88	78 71	37 18	48 64	06 57
22 15	78 15	69 84	32 52	32 54	15 12	54 02	01 37	38 37	12 93
93 29	12 18	27 30	30 55	91 87	50 57	58 51	49 36	12 53	96 40
45 04	77 97	36 14	99 45	52 95	69 85	03 83	51 87	85 56	22 37
44 91	99 49	89 39	94 60	48 49	06 77	64 72	59 26	08 51	25 57
16 23	91 02	19 96	47 59	89 65	27 84	30 92	63 37	26 24	23 66
04 50	65 04	65 65	82 42	70 51	55 04	61 47	88 83	99 34	82 37
32 70	17 72	03 61	66 26	24 71	22 77	88 33	17 78	08 92	73 49
03 64	59 07	42 95	81 39	06 41	20 81	92 34	51 90	39 08	21 42
62 49	00 90	67 86	93 48	31 83	19 07	67 68	49 03	27 47	52 03
61 00	95 86	98 36	14 03	48 88	51 07	33 40	06 86	33 76	68 57
89 03	90 49	28 74	21 04	09 96	60 45	22 03	52 80	01 79	33 18
01 72	33 85	52 40	60 07	06 71	89 27	14 29	55 24	85 79	31 96
27 56	49 79	34 34	32 22	60 53	91 17	33 26	44 70	93 14	99 70
49 05	74 48	10 55	35 25	24 28	20 22	35 66	66 34	26 35	91 23
49 74	37 25	97 26	33 94	42 23	01 28	59 58	92 69	03 66	73 82
20 26	22 43	88 08	19 85	08 12	47 65	65 63	56 07	97 85	56 79
48 87	77 96	43 39	76 93	08 79	22 18	54 55	93 75	97 26	90 77
08 72	87 46	75 73	00 11	27 07	05 20	30 85	22 21	04 67	19 13
95 97	98 62	17 27	31 42	64 71	46 22	32 75	19 32	20 99	94 85
37 99	57 31	70 40	46 55	46 12	24 32	36 74	69 20	72 10	95 93
05 79	58 37	85 33	75 18	88 71	23 44	54 28	00 48	96 23	66 45
55 85	63 42	00 79	91 22	29 01	41 39	51 40	36 65	26 11	78 32
67 28	96 25	68 36	24 72	03 85	49 24	05 69	64 86	08 19	91 21
85 86	94 78	32 59	51 82	86 43	73 84	45 60	89 57	06 87	08 15
40 10	60 09	05 88	78 44	63 13	58 25	37 11	18 47	75 62	52 21
94 55	89 48	90 80	77 80	26 89	87 44	23 74	66 20	20 19	26 52
11 63	77 77	23 20	33 62	62 19	29 03	94 15	56 37	14 09	47 16
64 00	26 04	54 55	38 57	94 62	68 40	26 04	24 25	03 61	01 20
50 94	13 23	78 41	60 58	10 60	88 46	30 21	45 98	70 96	36 89
66 98	37 96	44 13	45 05	34 59	75 85	48 97	27 19	17 85	48 51
66 91	42 83	60 77	90 91	60 90	79 62	57 66	72 28	08 70	96 03
33 58	12 18	02 07	19 40	21 29	39 45	90 42	58 84	85 43	95 67
52 49	40 16	72 40	73 05	50 90	02 04	98 24	05 30	27 25	20 88
74 98	93 99	78 30	79 47	96 92	45 58	40 37	89 76	84 41	74 68
50 26	54 30	01 88	69 57	54 45	69 88	23 21	05 69	93 44	05 32
49 46	61 89	33 79	96 84	28 34	19 35	28 73	39 59	56 34	97 07
19 65	13 44	78 39	73 88	62 03	36 00	25 96	86 76	67 90	21 68
64 17	47 67	87 59	81 40	72 61	14 00	28 28	55 86	23 38	16 15
18 43	97 37	68 97	56 56	57 95	01 88	11 89	48 07	42 60	11 92
65 58	60 87	51 09	96 61	15 53	66 81	66 88	44 75	37 01	28 88
79 90	31 00	91 14	85 65	31 75	43 15	45 93	64 78	34 53	88 02
07 23	00 15	59 05	16 09	94 42	20 40	63 76	65 67	34 11	94 10
90 08	14 24	01 51	95 46	30 32	33 19	00 14	19 28	40 51	92 69
53 82	62 02	21 82	34 13	41 03	12 85	65 30	00 97	56 30	15 48
98 17	26 15	04 50	76 25	20 33	54 84	39 31	23 33	59 64	96 27
08 91	12 44	82 40	30 62	45 50	64 54	65 17	89 25	59 44	99 95
37 21	46 77	84 87	67 39	85 54	97 37	33 41	11 74	90 50	29 62

Taken from Kendall and Babington-Smith's collection in *Tracts for Computers*, published by the Department of Statistics and Computer Science, University College, London and by permission of the publishers.

APPENDIX 3

Distribution of Chi-squared

Degrees of freedom	Probability, p											
	0.99	0.98	0.95	0.90	0.80	0.50	0.20	0.10	0.05	0.02	0.01	0.001
1	0.000	0.001	0.004	0.016	0.064	0.455	1.64	2.71	3.84	5.41	6.64	10.83
2	0.020	0.040	0.103	0.211	0.446	1.386	3.22	4.61	5.99	7.82	9.21	13.82
3	0.115	0.185	0.352	0.584	1.005	2.366	4.64	6.25	7.82	9.84	11.35	16.27
4	0.297	0.429	0.711	1.064	1.649	3.357	5.99	7.78	9.49	11.67	13.28	18.47
5	0.554	0.752	1.145	1.610	2.343	4.351	7.29	9.24	11.07	13.39	15.09	20.52
6	0.872	1.134	1.635	2.204	3.070	5.35	8.56	10.65	12.59	15.03	16.81	22.46
7	1.239	1.564	2.167	2.833	3.822	6.35	9.80	12.02	14.07	16.62	18.48	24.32
8	1.646	2.032	2.733	3.490	4.594	7.34	11.03	13.36	15.51	18.17	20.09	26.13
9	2.088	2.532	3.325	4.168	5.380	8.34	12.24	14.68	16.92	19.68	21.67	27.88
10	2.558	3.059	3.940	4.865	6.179	9.34	13.44	15.99	18.31	21.16	23.21	29.59
11	3.05	3.61	4.58	5.58	6.99	10.34	14.63	17.28	19.68	22.62	24.73	31.26
12	3.57	4.18	5.23	6.30	7.81	11.34	15.81	18.55	21.03	24.05	26.22	32.91
13	4.11	4.77	5.89	7.04	8.63	12.34	16.99	19.81	22.36	25.47	27.69	34.53
14	4.66	5.37	6.57	7.79	9.47	13.34	18.15	21.06	23.69	26.87	29.14	36.12
15	5.23	5.99	7.26	8.55	10.31	14.34	19.31	22.31	25.00	28.26	30.58	37.70
16	5.81	6.61	7.96	9.31	11.15	15.34	20.47	23.54	26.30	29.63	32.00	39.25
17	6.41	7.26	8.67	10.09	12.00	16.34	21.62	24.77	27.59	31.00	33.41	40.79
18	7.02	7.91	9.39	10.87	12.86	17.34	22.76	25.99	28.87	32.35	34.81	42.31
19	7.63	8.57	10.12	11.65	13.72	18.34	23.90	27.20	30.14	33.69	36.19	43.82
20	8.26	9.24	10.85	12.44	14.58	19.34	25.04	28.41	31.41	35.02	37.57	45.32
21	8.90	9.92	11.59	13.24	15.45	20.34	26.17	29.62	32.67	36.34	38.93	46.80
22	9.54	10.60	12.34	14.04	16.31	21.34	27.30	30.81	33.92	37.66	40.29	48.27
23	10.20	11.29	13.09	14.85	17.19	22.34	28.43	32.01	35.17	38.97	41.64	49.73
24	10.86	11.99	13.85	15.66	18.06	23.34	29.55	33.20	36.42	40.27	42.98	51.18
25	11.52	12.70	14.61	16.47	18.94	24.34	30.68	34.38	37.65	41.57	44.31	52.62
26	12.20	13.41	15.38	17.29	19.82	25.34	31.80	35.66	38.89	42.86	45.64	54.05
27	12.88	14.13	16.15	18.11	20.70	26.34	32.91	36.74	40.11	44.14	46.96	55.48
28	13.57	14.85	16.93	18.94	21.59	27.34	34.03	37.92	41.34	45.42	48.28	56.89
29	14.26	15.57	17.71	19.77	22.48	28.34	35.14	39.09	42.56	46.69	49.59	58.30
30	14.95	16.31	18.49	20.60	23.36	29.34	36.25	40.26	43.77	47.96	50.89	59.70

Taken from Fisher and Yates: *Statistical Tables for Biological, Agricultural and Medical Research*, published by Longman Group Ltd., London, (previously published by Oliver and Boyd, Edinburgh), and by permission of the authors and publishers.

APPENDIX 4

Percentage points of Student's t distribution

P	25	10	5	2	1	0.2	0.1
1	2.41	6.31	12.71	31.82	63.66	318.3	636.6
2	1.60	2.92	4.30	6.96	9.92	22.33	31.60
3	1.42	2.35	3.18	4.54	5.84	10.21	12.92
4	1.34	2.13	2.78	3.75	4.60	7.17	8.61
5	1.30	2.02	2.57	3.36	4.03	5.89	6.87
6	1.27	1.94	2.45	3.14	3.71	5.21	5.96
7	1.25	1.89	2.36	3.00	3.50	4.79	5.41
8	1.24	1.86	2.31	2.90	3.36	4.50	5.04
9	1.23	1.83	2.26	2.82	3.25	4.30	4.78
10	1.22	1.81	2.23	2.76	3.17	4.14	4.59
12	1.21	1.78	2.18	2.68	3.05	3.93	4.32
15	1.20	1.75	2.13	2.60	2.95	3.73	4.07
20	1.18	1.72	2.09	2.53	2.85	3.55	3.85
24	1.18	1.71	2.06	2.49	2.80	3.47	3.75
30	1.17	1.70	2.04	2.46	2.75	3.39	3.65
40	1.17	1.68	2.02	2.42	2.70	3.31	3.55
60	1.16	1.67	2.00	2.39	2.66	3.23	3.46
120	1.16	1.66	1.98	2.36	2.62	3.16	3.37
∞	1.15	1.64	1.96	2.33	2.58	3.09	3.29

(Degrees of freedom shown in leftmost column)

Based on the *Biometrika* tables and by permission of the Trustees of Biometrika.

APPENDIX 5

Village survey questionnaire

NAME OF VILLAGE............ NUMBER......

1) Where do you buy or receive:
   - *a*) groceries
   - *b*) fish
   - *c*) bread
   - *d*) meat
   - *e*) fuel

2) Where do you shop for:
   - *a*) hardware
   - *b*) shoes
   - *c*) adult clothes
   - *d*) children's clothes
   - *e*) furniture
   - *f*) electrical goods (T.V.)

3*a*) Where do you obtain these services?
   - *a*) dentist
   - *b*) doctor
   - *c*) optician
   - *d*) chemist
   - *e*) hairdressing
   - *f*) car service/repair/M.O.T.

3*b*) Where do you obtain these services?
   - *a*) solicitor
   - *b*) building society
   - *c*) vet
   - *d*) bank

4) Where do you go for:
   - *a*) cinema
   - *b*) theatre
   - *c*) club
   - *d*) other entertainments

5*a*) Which secondary schools do the older children attend?

5*b*) Where do working members of the household work?
   - *a*) head
   - *b*) wife (*if working*)
   - *c*) sons (*if any*)
   - *d*) daughters (*if any*)

6) Where is your newsagent?

REMARKS:

APPENDIX 6

Farm enquiry

1) *Location of farm:*

        Grid reference......

2) *Size of holding:*

        ...... acres   Consolidated/fragmented

3) *Crops:*

Acres		Acres		Acres
Wheat		Mangolds		Potatoes
Barley		Kale		Vegetables
Oats		Turnips and Swedes		Small fruit
Rye		Permanent grass		Orchard
Rape		Temporary grass		Others

4) *Livestock:*

Number		Number		Number
Dairy cattle		Bulls		Sheep
Dairy heifers		Other cattle		Lambs
Beef cattle		Pigs		Poultry

5) *Markets:*

6) *Labour:*

        Full time...... Part time...... Seasonal .....